Design of Roadside Channels with Flexible Linings

Technical Report Documentation Page

1. Report No. FHWA-NHI-05-114 HEC 15	2. Government Accession No.	3. Recipient's Catalog No.
4. Title and Subtitle Design of Roadside Channels with Flexible Linings Hydraulic Engineering Circular Number 15, Third Edition		5. Report Date September 2005
		6. Performing Organization Code
7. Author(s) Roger T. Kilgore and George K. Cotton		8. Performing Organization Report No.
9. Performing Organization Name and Address Kilgore Consulting and Management 2963 Ash Street Denver, CO 80207		10. Work Unit No. (TRAIS)
		11. Contract or Grant No. DTFH61-02-D-63009/T-63044
12. Sponsoring Agency Name and Address Federal Highway Administration National Highway Institute 4600 North Fairfax Drive Suite 800 Arlington, Virginia 22203 Office of Bridge Technology 400 Seventh Street Room 3202 Washington D.C. 20590		13. Type of Report and Period Covered Final Report (3rd Edition) April 2004 – August 2005
		14. Sponsoring Agency Code
15. Supplementary Notes Project Manager: Dan Ghere – FHWA Resource Center Technical Assistance: Jorge Pagan, Joe Krolak, Brian Beucler, Sterling Jones, Philip L. Thompson (consultant)		

16. Abstract

Flexible linings provide a means of stabilizing roadside channels. Flexible linings are able to conform to changes in channel shape while maintaining overall lining integrity. Long-term flexible linings such as riprap, gravel, or vegetation (reinforced with synthetic mats or unreinforced) are suitable for a range of hydraulic conditions. Unreinforced vegetation and many transitional and temporary linings are suited to hydraulic conditions with moderate shear stresses.

Design procedures are given for four major categories of flexible lining: vegetative linings; manufactured linings (RECPs); riprap, cobble, gravel linings; and gabion mattress linings. Design procedures for composite linings, bends, and steep slopes are also provided. The design procedures are based on the concept of maximum permissible tractive force. Methods for determination of hydraulic resistance applied shear stress as well as permissible shear stress for individual linings and lining types are presented.

This edition includes updated methodologies for vegetated and manufactured lining design that addresses the wide range of commercial products now on the market. This edition also includes a unified design approach for riprap integrating alternative methods for estimating hydraulic resistance and the steep slope procedures. Other minor updates and corrections have been made. This edition has been prepared using dual units.

17. Key Word channel lining, channel stabilization, tractive force, resistance, permissible shear stress, vegetation, riprap, manufactured linings, RECP, gabions		18. Distribution Statement This document is available to the public from the National Technical Information Service, Springfield, Virginia, 22151	
19. Security Classif. (of this report) Unclassified	20. Security Classif. (of this page) Unclassified	21. No. of Pages 153	22. Price

Form DOT F 1700.7 (8-72)   Reproduction of completed page authorized

ACKNOWLEDGMENTS

First Edition

Mr. Jerome M. Normann of Federal Highway Administration wrote the first edition of this Hydraulic Engineering Circular. FHWA reviewers included Frank Johnson, Dennis Richards and Albert Lowe of the Hydraulics Branch. The manual was dated October 1975.

Second Edition

Dr. Y. H. Chen and Mr. G. K. Cotton of Simons, Li & Associates wrote the second edition of this Hydraulic Engineering Circular. It was published as report number FHWA-IP-87-7 dated April 1988 under contract number DTFH61-84-C-00055. The FHWA project managers were John M. Kurdziel and Thomas Krylowski. Philip L. Thompson, Dennis L. Richards, and J. Sterling Jones were FHWA technical assistants

Third Edition

Mr. Roger T. Kilgore and Mr. George K. Cotton wrote this third edition of this Hydraulic Engineering Circular. The authors appreciate guidance of FHWA technical project manager, Mr. Dan Ghere and the technical review comments of Jorge Pagan, Joe Krolak, Brian Beucler, Sterling Jones, and Philip Thompson.

TABLE OF CONTENTS

LIST OF TABLES

LIST OF FIGURES

LIST OF SYMBOLS

A	=	Cross-sectional area of flow prism, m^2 (ft^2)
AOS	=	Measure of the largest effective opening in an engineering fabric, as measured by the size of a glass bead where five percent or less by weight will pass through the fabric
B	=	Bottom width of trapezoidal channel, m (ft)
C_f	=	cover factor
CG	=	Channel Geometry
D_{50}	=	Particle size of gradation, of which 50 percent of the mixture is finer by weight, m (ft)
D_{85}	=	Particle size of gradation, of which 85 percent of the mixture is finer by weight, m (ft)
d	=	Depth of flow in channel for the design flow, m (ft)
d_a	=	Average depth of flow in channel, m (ft)
Δd	=	Change in depth due to super elevation of flow in a bend, m (ft)
d_n	=	Depth of normal or uniform flow, m (ft)
F_d	=	Drag force in direction of flow
F_L	=	Lift force
Fr	=	Froude number, ratio of inertial forces to gravitational force in a system
g	=	gravitational acceleration, m/s^2 (ft/s^2)
h	=	Average height of vegetation, mm (in)
K_b	=	ratio of maximum shear stress in bend to maximum shear stress upstream from bend
K_1	=	ratio of channel side shear to bottom shear stress
K_2	=	tractive force ratio
k_s	=	roughness height, mm (in)
ℓ	=	Moment arms of forces acting on riprap in a channel
L_p	=	protected length downstream from bend, m (ft)
MEI	=	Stiffness factor, $N \cdot m^2$ ($lb \cdot ft^2$)
n	=	Manning's roughness coefficient
n_e	=	composite channel lining equivalent Manning's n
P	=	Wetted perimeter of flow prism, m (ft)
P_L	=	Wetted perimeter of low-flow channel, m (ft)
PC	=	Point on curve
PT	=	Point on tangent
Q	=	Discharge, flow rate, m^3/s (ft^3/s)
R	=	Hydraulic radius, A/P, m (ft)
R_C	=	Mean radius of channel center line, m (ft)
REG	=	Roughness element geometry
S_o	=	Average channel gradient
S_f	=	Energy (friction) gradient
SF	=	Safety factor
S_{50}	=	Mean value of the short axis lengths of the roughness element, m (ft)

T	=	Channel top width (water surface), m (ft)
V	=	Mean channel velocity, m/s (ft/s)
V_*	=	Shear velocity, m/s (ft/s)
W_S	=	Weight of riprap element, N (lb)
Y_{50}	=	Mean value of the average of the long and median axes of the roughness element, m (ft)
Z	=	Side slope; cotangent of angle measured from horizontal, $Z = \tan^{-1}\theta$
α	=	Unit conversion constant for SI and CU; equation specific
α_c	=	Angle of channel bottom slope
β	=	Angle between weight vector and the resultant in the plane of the side slope
γ	=	Unit weight of water, N/m^3 (lb/ft^3)
δ	=	Angle between the drag vector and resultant in the plane of the side slope
θ	=	Angle of side slope (bank) measured from horizontal
ϕ	=	Angle of repose of coarse, noncohesive material, degrees
η	=	Stability number
η'	=	Stability number for side slopes
σ	=	Bed material gradation
τ_b	=	Shear stress in a bend, N/m^2 (lb/ft^2)
τ_d	=	Shear stress in channel at maximum depth, d, N/m^2 (lb/ft^2)
τ_l	=	Shear stress on a RECP that results in 12.5 mm (0.5 in) of erosion in 30 minutes
τ_o	=	Mean boundary shear stress, N/m^2 (lb/ft^2)
τ_p	=	Permissible shear stress, N/m^2 (lb/ft^2)
τ_s	=	Shear stress on sides of channel, N/m^2 (lb/ft^2)

GLOSSARY

Angle of Repose: Angle of slope formed by particulate material under the critical equilibrium condition of incipient motion.

Apparent Opening Size (AOS): Measure of the largest effective opening in an engineering fabric, as measured by the size of a glass bead where five percent or less by weight will pass through the fabric (formerly called the equivalent opening size, EOS).

Compaction: The closing of pore spaces among the particles of soil and rock, generally caused by running heavy equipment over the soil during construction.

Customary Units (CU): Foot-pound system of units often referred to as English units.

Depth of Flow: Vertical distance from the bottom of a channel to the water surface, also referred to as the maximum depth of flow.

Design Discharge: Discharge at a specific location defined by an appropriate return period to be used for design purposes.

Engineering Fabric: Permeable textile (or filter fabric) used below riprap to prevent piping and permit natural seepage to occur.

Erosion Control Blanket (ECB): A degradable material, composed primarily of processed natural organic materials, manufactured or fabricated into rolls designed to reduce soil erosion and assist in the growth, establishment and protection of vegetation.

Filter Blanket: One or more layers of graded noncohesive material placed below riprap to prevent soil piping and permit natural seepage to occur.

Freeboard: Vertical distance from the water surface to the top of the channel at design condition.

Gabion: Compartmented rectangular containers made of galvanized steel hexagonal wire mesh and filled with stone.

Hydraulic Radius: Flow area divided by wetted perimeter.

Hydraulic Resistance: Resistance encountered by water as it moves through a channel, commonly described by Manning's n.

Hydrostatic Pressure: Pressure exerted at a depth below the water surface for flow at constant velocity or at rest.

Incipient Motion: Conditions at that point in time when any increase in factors responsible for particle movement causes motion.

Lining, Composite: Combination of lining materials in a given cross section (e.g., riprap in low-flow channel and vegetated side slopes).

Lining, Flexible: Lining material with the capacity to adjust to settlement typically constructed of a porous material that allows infiltration and exfiltration.

Lining, Long-term: Lining designed for long-term use. Although many flexible linings do have limited functional life spans, their durability is compatible with the service life of the drainageway.

Lining, Rigid: Lining material with no capacity to adjust to settlement constructed of nonporous material with smooth finish that provides a large conveyance capacity (e.g. concrete, soil cement).

Lining, Temporary: Lining designed for an interim condition, typically serving the needs of construction sequencing. Temporary linings will be removed.

Lining, Transitional: Lining designed for an interim condition, typically to assist in development of a permanent vegetative lining. Transitional linings will not be removed.

Normal Depth: Depth of a uniform channel flow.

Open Weave Textile (OWT): A temporary degradable ECB composed of natural or polymer yarns woven into a matrix used to provide erosion control and facilitate vegetation establishment.

Permeability: Property of a soil that enables water or air to move through it.

Retardance Classification: Qualitative description of the resistance to flow offered by various types of vegetation.

Riprap: Broken rock, cobbles, or boulders placed on side slopes or in channels for protection against the action of water.

Rolled Erosion Control Product (RECP): A temporary degradable or long-term non-degradable material manufactured or fabricated into rolls designed to reduce soil erosion and assist in the growth, establishment, and protection of vegetation.

Rundown: Steep, generally short, conveyance channel used adjacent to bridge abutments or other embankment locations.

Roadside Channel: Stabilized drainageway used to collect water from the roadway and adjacent areas and to deliver it to an inlet or main drainageway.

Shear Stress: Stress developed on the wetted area of the channel for a given hydraulic conditions that acts in the direction of the flow; stress is force per unit wetted area.

Shear Stress, Permissible: Force at which the channel lining will fail.

Side Slope: Slope of the sides of a channel defined as the run corresponding to a unit rise; represented by Z as in 1:Z (vertical:horizontal).

Superelevation: Local increase in water surface on the outside of a bend.

System International (SI): Meter kilogram second system of units often referred to as metric units.

Tractive Force: Force developed due to the shear stress acting on the perimeter of a channel section that acts in the direction of flow on the channel bottom; equals the shear stress on the channel section multiplied by the wetted channel area.

Turf Reinforcement Mat (TRM): A non-degradable RECP composed of UV stabilized synthetic fibers, filaments, netting and/or wire mesh processed into a three-dimensional matrix. TRMs provide sufficient thickness, strength and void space to permit soil filling and establishment of grass roots within the matrix.

Uniform flow: The flow condition where the rate of head loss due to friction is equal to bottom slope of the channel, that is, $S_f = S_o$, where S_f is the friction slope and S_o is the bottom slope.

Velocity, Mean: Discharge divided by the area of flow.

Velocity, Permissible: Mean velocity that will not cause serious erosion of the channel.

CHAPTER 1: INTRODUCTION

This manual addresses the design of small open channels called roadside channels that are constructed as part of a highway drainage system. Roadside channels play an important role in the highway drainage system as the initial conveyance for highway runoff. Roadside channels are often included as part of the typical roadway section. Therefore, the geometry of roadside channels depends on available right-of-way, flow capacity requirements, and the alignment and profile of the highway. The procedures in this manual may also be used for ancillary roadside drainage features such as rundowns.

Roadside channels capture sheet flow from the highway pavement and backslope and convey that runoff to larger channels or culverts within the drainage system. This initial concentration of runoff may create hydraulic conditions that are erosive to the soil that forms the channel boundary. To perform reliably, the roadside channel is often stabilized against erosion by placing a protective lining over the soil. This manual presents a class of channel linings called flexible linings that are well suited for construction of small roadside channels.

This manual is presented in dual units. The SI (metric) units precede the customary units (CU) when units are given. Design examples are provided in both systems of units.

1.1 SCOPE AND APPLICABILITY

Channel lining materials fall into two classes: rigid or flexible channel linings. From an erosion control standpoint, the primary difference between rigid and flexible channel linings is their response to changes in channel shape (i.e. the width, depth and alignment). Flexible linings are able to adjust to some change in channel shape while rigid linings cannot. The ability to sustain some change in channel shape improves the overall integrity of the channel lining and reduces maintenance. Movement of a rigid lining at one location can result in a successive failure of the lining. Channel lining materials often experience forces such as frost heave, slumping or swelling of the underlying soils that can change the shape of the lining. These forces can displace rigid linings whereas flexible linings, if properly designed, will retain erosion-control capabilities.

Flexible linings also have several other advantages compared to rigid linings. They are generally less expensive, permit infiltration and exfiltration and can be vegetated to have a natural appearance. Flow in channels with flexible linings is similar to that found in natural small channels. More natural behavior offers better habitat opportunities for local flora and fauna. In many cases, flexible linings are designed to provide only transitional protection against erosion while vegetation establishes and becomes the permanent lining of the channel. Vegetative channel lining is also recognized as a best management practice for storm water quality design in highway drainage systems. The slower flow of a vegetated channel helps to deposit highway runoff contaminants (particularly suspended sediments) before they leave the highway right of way and enter streams.

Flexible linings have a limited hydraulic performance range (depth, grade, velocity and discharge). The magnitude of hydraulic force they can sustain without damage is limited by a number of factors including soil properties and roadway grading. Because of these limitations, flexible channel designs using the same lining material will vary from site to site and between regions of the country. Since the performance range for rigid channels is higher, such channels may be needed in cases where channel width is limited by right of way, but sufficient space exists for a high capacity channel.

Design procedures covered in this manual relate to flexible channel linings. Rigid linings are discussed only briefly so that the reader remains familiar with the full range of channel lining alternatives. The primary reference for the design of rigid channels is Hydraulic Design Series No. 4 "Introduction to Highway Hydraulics" (FHWA, 2001). For channels which require other protection measures, the design of energy dissipaters and grade-control structures can be found in Hydraulic Engineering Circular (HEC) No. 14 (FHWA, 1983).

Riprap design procedures covered in this manual are for prismatic channels typically having a maximum depth of 1.5 m (5 ft). However, the procedures for riprap design are not limited by depth with the exception of the limits cited on techniques for estimating Manning's roughness. The use of the procedures in Hydraulic Engineering Circular (HEC) No. 11 (FHWA, 1987) is recommended for nonprismatic channels.

The permissible tractive force and Manning's n values provided in this manual for grass-lined channels is based on the relative roughness theory, the biomechanical properties of grass (height, stiffness and density of the grass cover), and the properties of the underlying soil (particle size, density and plasticity). This method is comparable to methods used in agricultural channel design (USDA, 1987), but offers the highway designer more flexibility. This document provides a method of estimating grass properties for complex seed mix designs using a simple field test.

The current performance information for manufactured channel linings is based on industry testing and design recommendations. Product testing is routinely conducted by major manufacturers using either their own hydraulic laboratories (Clopper, Cabalka, Johnson, 1998) or using facilities at university labs. Industry protocols have been developed for large scale testing (ASTM D 6460) that provides a consistent test method for flexible channel lining materials. Small-scale tests (i.e. bench tests) have been developed that are intended for qualitative comparison of products and product quality verification. Data from bench testing is not sufficient to characterize the hydraulic performance of manufactured linings. While there is a qualitative understanding about manufactured-lining performance, large-scale testing is currently needed to determine performance properties.

1.2 BACKGROUND

Considerable development and research has been performed on rigid and flexible channel linings. Prior to the late 1960's, natural materials were predominantly used to stabilize channels. Typical materials included rock riprap, stone masonry, concrete, and vegetation. Since that time a wide variety of manufactured and synthetic channel linings applicable to both permanent and transitional channel stabilization have been introduced. Since the publication of the 1988 edition of HEC No. 15, erosion control material manufacturers have developed protocols for testing flexible linings in hydraulic laboratory flumes under controlled conditions.

The market for flexible channel lining products has expanded and there are a large number of channel stabilization materials currently available. Channel stabilization materials can be broadly classified based on their type and duration of installation. Two basic types of lining classes are defined: rigid and flexible. Rigid lining systems are permanent, long-duration installations. Flexible linings systems can either be long-term, transitional, or temporary installations. The following are examples of lining materials in each classification.

1. Rigid Linings:

    a. Cast-in-place concrete or asphaltic concrete

    b. Stone masonry and interlocking modular block

    c. Soil cement and roller compacted concrete

    d. Fabric form-work systems for concrete

    e. Partially grouted riprap

2. Flexible linings

    a. Long-term

        i. Vegetative (typically grass species)

        ii. Cobbles

        iii. Rock Riprap

        iv. Wire-enclosed riprap (gabions)

        v. Turf reinforcement (non-degradable)

    b. Transitional

        i. Bare soil

        ii. Vegetative (annual grasses)

        iii. Gravel mulch

        iv. Open-weave textile (degradable)

        v. Erosion control blankets (degradable)

        vi. Turf reinforcement (non-degradable)

    c. Temporary

        i. Bare soil

        ii. Vegetative (annual grasses)

        iii. Gravel mulch

        iv. Open-weave textile (degradable)

        v. Erosion control blankets (degradable)

Sprayed on mulch is a common application for erosion control on hill slopes. Mulch is combined with a glue or tackifier to form slurry that is pumped at high pressure onto the hill slope. The only channel lining tested in this class is fiberglass roving (McWhorter, Carpenter and Clark, 1968). This lining is not in use because during maintenance operations, mowers can rip up large sections of the roving. Also, although some tackifiers have been reported to encourage growth, asphalt tackifier usually inhibits vegetation establishment and growth.

An emerging product in this class is a form of sprayed on composting. Used both for hill slopes and for channels, the objective of the product is to accelerate vegetative establishment. As such, composting does not represent a lining product class, but is a strategy to shorten transition periods. Other new products may emerge in this class, but until full scale testing is conducted (in accordance with ASTM D 6460) they will not be covered in this manual. Products that address only hill slope or embankment erosion control and not channel applications are also not included in this manual.

1.3 RIGID LININGS

Rigid linings (Figure 1.1) are useful in flow zones where high shear stress or non-uniform flow conditions exist, such as at transitions in channel shape or at an energy dissipation structure. They can be designed to include an impermeable membrane for channels where loss of water from seepage is undesirable. Since rigid linings are non-erodible the designer can use any channel shape that is necessary to convey the flow and provide adequate freeboard. This may be necessary where right-of-way constrains the channel width.

Figure 1.1. Rigid Concrete Channel Lining

Despite the non-erodible nature of rigid linings, they are susceptible to failure from foundation instability. The major cause of failure is undermining that can occur in a number of ways. Inadequate erosion protection at the outfall, at the channel edges, and on bends can initiate undermining by allowing water to carry away the foundation material and leaving the channel to break apart. Rigid linings may also break up and deteriorate due to conditions such as a high water table or swelling soils that exert an uplift pressure on the lining. Once a rigid lining is locally broken and displaced upward, the lining continues to move due to dynamic uplift and drag forces. The broken lining typically forms large, flat slabs that are particularly susceptible to these forces. Freeze thaw cycles may also stress rigid channels. The repeated cycling of these forces can cause fine particles to migrate within the underlying soil causing filter layers and weep holes to clog and further increase uplift pressure on the lining.

Rigid linings are particularly vulnerable to a seasonal rise in water table that can cause a static uplift pressure on the lining. If a rigid lining is needed in such conditions, a reliable system of under drains and weep holes should be a part of the channel design. The migration of soil fines into filter layers should be evaluated to ensure that the ground water is discharged without filter clogging or collapse of the underlying soil. A related case is the build up of soil pore pressure behind the lining when the flow depth in the channel drops quickly. Use of watertight joints and backflow preventers on weep holes can help to reduce the build up of water behind the lining.

Construction of rigid linings requires specialized equipment and costly materials. As a result, the cost of rigid channel linings is typically higher than an equivalent flexible channel lining. Prefabricated linings can be a less expensive alternative if shipping distances are not excessive. Many highway construction projects include paving materials (concrete and asphaltic concrete) that are also used in rigid channel linings. This may provide an economy of scale when similar materials are used for both paving and channel construction.

1.4   FLEXIBLE LININGS

Flexible linings can meet a variety of design objectives and serve a variety of roles in the construction of a project where prismatic channels are required for conveying stormwater runoff. Flexible channel linings are best suited to conditions of uniform flow and moderate shear stresses. Channel reaches with accelerating or decelerating flow (expansions, contractions, drops and backwater) and waves (transitions, flows near critical depth, and shorelines) will require special analysis and may not be suitable for flexible channel linings.

Several terms are used to describe the longevity of flexible linings - permanent, long-term, transitional, temporary, and short-term – to name a few. Recognizing that nothing is permanent, long-term is defined as serving the desired purpose throughout the lifetime of the drainage channel given appropriate maintenance. The other terms imply that changes must occur either in the removal of the channel or replacement of one lining type with another. However, the designer should keep in mind not only the manufacturer's claims of longevity, but also site-specific maintenance practices and climate or geographic location in selecting a lining type for a given transitional or temporary application.

1.4.1   Long-term Flexible Linings

Long-term flexible linings are used where roadside channels require protection against erosion for the service life of the channel.

1.4.1.1   Vegetation

Vegetative linings consist of seeded or sodded grasses placed in and along the channel (Figure 1.2). Grasses are seeded and fertilized according to the requirements of that particular variety or mixture. Sod is laid with the longest side parallel to the flow direction and should be secured with pins or staples.

Figure 1.2. Vegetative Channel Lining

Vegetation is one of the most common long-term channel linings. Most roadside channels capture only initial highway runoff and so remain dry most of the time. For these conditions, upland species of vegetation (typically grass) provide a good lining. However, upland species of vegetation are not suited to sustained flow conditions or long periods of submergence. Common design practice for vegetative channels with sustained low flow and intermittent high flows is to provide a composite lining with riprap or concrete providing a low flow section. There are plant species that are adapted to wet low land conditions that can be used for the low flow channel in cases that warrant the additional design and construction effort (wetland replacement for example).

Where vegetation provides the long-term channel lining, there is a transition period between seeding and vegetation establishment. The initial unvegetated condition of the lining is followed by a period of vegetation establishment that can take several growing seasons. The channel is vulnerable to erosion during the transition. Transitional flexible linings provide erosion protection during the vegetation establishment period. These linings are typically degradable and do not provide ongoing stabilization of the channel after vegetation is established. Non-degradable linings have an expected life of several years beyond vegetation establishment, which enhances the performance of the vegetation. At this time it is not known how long an installation of non-degradable flexible linings will last together with vegetation.

1.4.1.2 Cobble Lining

Cobble lining consists of stone in the size range of small cobbles, 64 to 130 mm (2.5 to 5 inches), and tends to have a uniform gradation. The cobble layer is placed on engineering fabric on a prepared grade (Figure 1.3). The cobble material is composed of uniformly graded, durable stone that is free of organic matter. Cobbles are typically alluvial in origin and smooth and rounded in appearance.

Cobble linings are often used when a decorative channel design is needed. Cobble linings are composed of smooth stones that do not interlock, so they are not suitable for placement on steep grades or on channel side slopes that are steep. As with riprap and gabion linings, a filter material is required between the stone and the underlying soil.

Figure 1.3. Cobble Channel Lining

1.4.1.3 Rock Riprap

Rock riprap is placed on a filter blanket or prepared slope to form a well-graded mass with a minimum of voids (Figure 1.4). Rocks should be hard, durable, preferably angular in shape, and free from overburden, shale, and organic material. The rock should be durable and resistance to disintegration from chemical and physical weathering. The performance of riprap should be determined from service records for a quarry or pit, or from specified field and laboratory tests.

Riprap and gabion linings can perform in the initial range of hydraulic conditions where rigid linings are used. Stones used for riprap and gabion installations preferably have an angular shape that allows stones to interlock. These linings usually require a filter material between the stone and the underlying soil to prevent soil washout. In most cases, an engineering fabric is used as the filter. Care should be taken to provide adequate permeability in the filter to prevent uplift pressures on the lining.

Figure 1.4. Riprap Channel Lining

1.4.1.4 Wire-Enclosed Riprap

Wire-enclosed riprap (gabions) is a wire container or enclosure structure that binds units of the riprap lining together. The wire enclosure normally consists of a rectangular container made of steel wire woven in a uniform pattern, and reinforced on corners and edges with heavier wire (Figure 1.5 and Figure 1.6). The containers are filled with stone, connected together, and anchored to the channel side slope. Stones must be well graded and durable. The forms of wire-enclosed riprap vary from thin mattresses to box-like gabions. Wire-enclosed riprap is typically used when rock riprap is either not available or not large enough to be stable. Although flexible, gabion movement is restricted by the wire mesh.

Figure 1.5. Wire-Enclosed Riprap

Figure 1.6. Installed Wire-Enclosed Riprap

1.4.1.5 Turf Reinforcement

Depending on the application, materials, and method of installation, turf reinforcement may serve a transitional or long-term function. The concept of turf reinforcement is to provide a structure to the soil/vegetation matrix that will both assist in the establishment of vegetation and provide support to mature vegetation. Two types of turf reinforcement are commonly available: gravel/soil methods and turf reinforcement mats (TRMs)

Soil/gravel turf reinforcement is to mix gravel mulch (see Section 1.4.2.2) into on-site soils and to seed the soil-gravel layer. The rock products industry provides a variety of uniformly graded gravels for use as mulch and soil stabilization. A gravel/soil mixture provides a non-degradable lining that is created as part of the soil preparation and is followed by seeding.

A TRM is a non-degradable RECP composed of UV stabilized synthetic fibers, filaments, netting and/or wire mesh processed into a three-dimensional matrix. TRMs provide sufficient thickness, strength and void space to permit soil filling and establishment of grass roots within the matrix. The mat, shown in Figure 1.7 and Figure 1.8, is laid parallel to the direction of flow. TRM is stiffer, thicker (minimum of 6 mm (0.25 in)) and denser than an erosion control blanket (ECB). These material properties improve erosion resistance. The TRM is secured with staples and anchored into cutoff trenches at intervals along the channel. Two methods of seeding can be used with TRM. One choice is to seed before placement of the TRM, which allows the plant stems to grow through the mat. The second choice is to first place the TRM then cover the mat with soil and then seed. This method allows the plant roots to grow within the mat.

Figure 1.7. TRM Profile with Vegetation/Soil/TRM Matrix (Source: ECTC)

Figure 1.8. Installed TRM Lining Before Vegetation (Source: ECTC)

1.4.2 Transitional and Temporary Flexible Linings

Transitional linings are intended and designed to facilitate establishment of the long-term flexible lining. Commonly the long-term lining would be vegetation. Temporary channel linings are used without vegetation to line channels that are part of construction site erosion control systems and other short-term channels. In some climates, rapidly growing annual grass species establish quickly enough to be used as a temporary channel lining.

Many of the transitional and temporary linings are described as degradable. Functionally, this means that the structural matrix of the lining breaks down as a result of biological processes and/or UV light exposure. In the case of organic materials, the lining becomes a natural part of the underlying soil matrix. In the case of degradable plastics, many products lose their structural integrity and degrade to a powder that remains in the soil. The long-term environmental effects of widespread use of such products are unknown and require study.

1.4.2.1 Bare Soil

The properties of site soils are important in the design of all flexible linings because erosion of the underlying soil is one of the main performance factors in lining design. The erodibility of soil is a function of texture, plasticity and density. Bare soil alone can be a sufficient lining in climates where vegetation establishes quickly and the interim risk of soil erosion is small. Bare-soil channels may have a low risk of erosion if grades are mild, flow depths are shallow, and soils have a high permissible shear stress resistance (high plasticity cohesive soils or gravelly non-cohesive soils).

1.4.2.2 Gravel Mulch

Gravel mulch is a non-degradable erosion control product that is composed of coarse to very coarse gravel, 16 mm to 64 mm (0.6 to 2.5 inch), similar to an AASHTO No. 3 coarse aggregate. Placement of gravel is usually done immediately after seeding operations. Gravel mulch is particularly useful on windy sites or where it is desirable to augment the soil with coarse particles. Application of gravel can reduce wheel rutting on shoulders and in ditches. It can also be used to provide a transition between riprap and soil. Unlike riprap and other stone linings, gravel mulch should be placed directly on the soil surface without an underlying filter fabric. Constructing intermediate cutoff trenches that are filled with gravel enhances stability of the lining.

1.4.2.3 Vegetation (Annual Grass)

If the construction phasing permits and the climate is suitable, annual grasses can be seeded in time to establish a transitional vegetative lining. Seed mixes typically include rapidly growing annual grasses. Soil amendments including the application of fertilizer and compost improve grass establishment. To be effective, these annual grasses need to be well established though the transition period and at a sufficient density to provide erosion control. Sodding is another rapid method of vegetation establishment for ditches. The sod needs to be staked to the ditch perimeter where flow is expected to prevent wash out.

1.4.2.4 Open-weave Textile (OWT)

Open-weave textiles are a degradable rolled erosion control product that is composed of natural or polymer yarns woven into a matrix. OWT can be used together with straw mulch to retain soil moisture and to increase the density and thickness of the lining. OWT is more flexible, thinner and less dense compared to erosion control blankets (ECB). The OWT (Figure 1.9 and Figure 1.10) is loosely laid in the channel parallel to the direction of flow. OWT is secured with staples and by placement of the fabric into cutoff trenches at intervals along the channel. Placement of OWT is usually done immediately after seeding operations.

Figure 1.9. Open Weave Textile Lining

Figure 1.10. Installed Open Weave Textile Channel Lining

1.4.2.5 Erosion control blanket (ECB)

Erosion control blanket is a degradable rolled erosion control product that is composed of an even distribution of natural or polymer fibers that are mechanically, structurally or chemically bound together to form a continuous mat (Figure 1.11). ECB is stiffer, thicker and denser than an open-weave textile (OWT). These material properties improve erosion resistance. The ECB is placed in the channel parallel to the direction of the flow and secured with staples and by placement of the blanket into cutoff trenches. When

ECBs are used and ultimately degrade, the long-term erosion protection is provided by the established vegetation.

Figure 1.11. Erosion Control Blanket (ECB) Lining (Source: ECTC)

This page intentionally left blank.

CHAPTER 2: DESIGN CONCEPTS

The design method presented in this circular is based on the concept of maximum permissible tractive force. The method has two parts, computation of the flow conditions for a given design discharge and determination of the degree of erosion protection required. The flow conditions are a function of the channel geometry, design discharge, channel roughness, channel alignment and channel slope. The erosion protection required can be determined by computing the shear stress on the channel lining (and underlying soil, if applicable) at the design discharge and comparing that stress to the permissible value for the type of lining/soil that makes up the channel boundary.

2.1   OPEN CHANNEL FLOW

2.1.1   Type of Flow

For design purposes in roadside channels, hydraulic conditions are usually assumed to be uniform and steady. This means that the energy slope is approximately equal to average ditch slope, and that the flow rate changes gradually over time. This allows the flow conditions to be estimated using a flow resistance equation to determine the so-called normal flow depth. Flow conditions can be either mild (subcritical) or steep (supercritical). Supercritical flow may create surface waves whose height approaches the depth of flow. For very steep channel gradients, the flow may splash and surge in a violent manner and special considerations for freeboard are required.

More technically, open-channel flow can be classified according to three general conditions:

- uniform or non-uniform flow
- steady or unsteady flow
- subcritical or supercritical flow.

In uniform flow, the depth and discharge remain constant along the channel. In steady flow, no change in discharge occurs over time. Most natural flows are unsteady and are described by runoff hydrographs. It can be assumed in most cases that the flow will vary gradually and can be described as steady, uniform flow for short periods of time. Subcritical flow is distinguished from supercritical flow by a dimensionless number called the Froude number (Fr), which is defined as the ratio of inertial forces to gravitational forces in the system. Subcritical flow (Fr < 1.0) is characterized as tranquil and has deeper, slower velocity flow. In a small channel, subcritical flow can be observed when a shallow wave moves in both the upstream and downstream direction. Supercritical flow (Fr > 1.0) is characterized as rapid and has shallow, high velocity flow.   At critical and supercritical flow, a shallow wave only moves in the downstream direction.

2.1.2   Normal Flow Depth

The condition of uniform flow in a channel at a known discharge is computed using the Manning's equation combined with the continuity equation:

$$Q = \frac{\alpha}{n} AR^{2/3} S_f^{1/2} \qquad (2.1)$$

where,

Q   = discharge, m^3/s (ft^3/s)

n   = Manning's roughness coefficient, dimensionless

A   = cross-sectional area, m^2 (ft^2)

R   = hydraulic radius, m (ft)

S_f = friction gradient, which for uniform flow conditions equals the channel bed gradient, S_o, m/m (ft/ft)

α = unit conversion constant, 1.0 (SI), 1.49 (CU)

The depth of uniform flow is solved by rearranging Equation 2.1 to the form given in Equation 2.2. This equation is solved by trial and error by varying the depth of flow until the left side of the equation is zero.

$$\frac{Q\,n}{\alpha \cdot S_f^{\frac{1}{2}}} - AR^{\frac{2}{3}} = 0 \qquad\qquad (2.2)$$

2.1.3   Resistance to Flow

For rigid channel lining types, Manning's roughness coefficient, n, is approximately constant. However, for very shallow flows the roughness coefficient will increase slightly. (Very shallow is defined where the height of the roughness is about one-tenth of the flow depth or more.)

For a riprap lining, the flow depth in small channels may be only a few times greater than the diameter of the mean riprap size. In this case, use of a constant n value is not acceptable and consideration of the shallow flow depth should be made by using a higher n value.

Tables 2.1 and 2.2 provide typical examples of n values of various lining materials. Table 2.1 summarizes linings for which the n value is dependent on flow depth as well as the specific properties of the material. Values for rolled erosion control products (RECPs) are presented to give a rough estimate of roughness for the three different classes of products. Although there is a wide range of RECPs available, jute net, curled wood mat, and synthetic mat are examples of open-weave textiles, erosion control blankets, and turf reinforcement mats, respectively. Chapter 5 contains more detail on roughness for RECPs.

Table 2.2 presents typical values for the stone linings: riprap, cobbles, and gravels. These are highly depth-dependent for roadside channel applications. More in-depth lining-specific information on roughness is provided in Chapter 6. Roughness guidance for vegetative and gabion mattress linings is in Chapters 4 and 7, respectively.

Table 2.1. Typical Roughness Coefficients for Selected Linings

Lining Category	Lining Type	Manning's n[1]		
		Maximum	**Typical**	**Minimum**
Rigid	Concrete	0.015	0.013	0.011
	Grouted Riprap	0.040	0.030	0.028
	Stone Masonry	0.042	0.032	0.030
	Soil Cement	0.025	0.022	0.020
	Asphalt	0.018	0.016	0.016
Unlined	Bare Soil[2]	0.025	0.020	0.016
	Rock Cut (smooth, uniform)	0.045	0.035	0.025
RECP	Open-weave textile	0.028	0.025	0.022
	Erosion control blankets	0.045	0.035	0.028
	Turf reinforcement mat	0.036	0.030	0.024

[1]Based on data from Kouwen, et al. (1980), Cox, et al. (1970), McWhorter, et al. (1968) and Thibodeaux (1968).
[2]Minimum value accounts for grain roughness. Typical and maximum values incorporate varying degrees of form roughness.

Table 2.2. Typical Roughness Coefficients for Riprap, Cobble, and Gravel Linings

Lining Category	Lining Type	Manning's n for Selected Flow Depths[1]		
		0.15 m (0.5 ft)	**0.50 m (1.6 ft)**	**1.0 m (3.3 ft)**
Gravel Mulch	D_{50} = 25 mm (1 in.)	0.040	0.033	0.031
	D_{50} = 50 mm (2 in.)	0.056	0.042	0.038
Cobbles	D_{50} = 0.10 m (0.33 ft)	--[2]	0.055	0.047
Rock Riprap	D_{50} = 0.15 m (0.5 ft)	--[2]	0.069	0.056
	D_{50} = 0.30 m (1.0 ft)	--[2]	--[2]	0.080

[1]Based on Equation 6.1 (Blodgett and McConaughy, 1985). Manning's n estimated assuming a trapezoidal channel with 1:3 side slopes and 0.6 m (2 ft) bottom width.
[2]Shallow relative depth (average depth to D_{50} ratio less than 1.5) requires use of Equation 6.2 (Bathurst, et al., 1981) and is slope-dependent. See Section 6.1.

2.2   SHEAR STRESS

2.2.1   Equilibrium Concepts

Most highway drainage channels cannot tolerate bank instability and possible lateral migration. Stable channel design concepts focus on evaluating and defining a channel configuration that will perform within acceptable limits of stability. Methods for evaluation and definition of a stable configuration depend on whether the channel boundaries can be viewed as:

- essentially rigid (static)
- movable (dynamic).

In the first case, stability is achieved when the material forming the channel boundary effectively resists the erosive forces of the flow. Under such conditions the channel bed and banks are in static equilibrium, remaining basically unchanged during all stages of flow. Principles of rigid boundary hydraulics can be applied to evaluate this type of system.

In a dynamic system, some change in the channel bed and/or banks is to be expected due to transport of the sediments that comprise the channel boundary. Stability in a dynamic system is attained when the incoming supply of sediment equals the sediment transport rate. This condition, where sediment supply equals sediment transport, is referred to as dynamic equilibrium. Although some detachment and transport of bed and/or bank sediments occurs, this does not preclude attainment of a channel configuration that is basically stable. A dynamic system can be considered stable so long as the net change does not exceed acceptable levels. Because of the need for reliability, static equilibrium conditions and use of linings to achieve a stable condition is usually preferable to using dynamic equilibrium concepts.

Two methods have been developed and are commonly applied to determine if a channel is stable in the sense that the boundaries are basically immobile (static equilibrium): 1) the permissible velocity approach and 2) the permissible tractive force (shear stress) approach. Under the permissible velocity approach the channel is assumed stable if the mean velocity is lower than the maximum permissible velocity. The tractive force (boundary shear stress) approach focuses on stresses developed at the interface between flowing water and materials forming the channel boundary. By Chow's definition, permissible tractive force is the maximum unit tractive force that will not cause serious erosion of channel bed material from a level channel bed (Chow, 1979).

Permissible velocity procedures were first developed around the 1920's. In the 1950's, permissible tractive force procedures became recognized, based on research investigations conducted by the U.S. Bureau of Reclamation. Procedures for design of vegetated channels using the permissible velocity approach were developed by the SCS and have remained in common use.

In spite of the empirical nature of permissible velocity approaches, the methodology has been employed to design numerous stable channels in the United States and throughout the world. However, considering actual physical processes occurring in open-channel flow, a more realistic model of detachment and erosion processes is based on permissible tractive force which is the method recommended in this publication.

2.2.2 Applied Shear Stress

The hydrodynamic force of water flowing in a channel is known as the tractive force. The basis for stable channel design with flexible lining materials is that flow-induced tractive force should not exceed the permissible or critical shear stress of the lining materials. In a uniform flow, the tractive force is equal to the effective component of the drag force acting on the body of water, parallel to the channel bottom (Chow, 1959). The mean boundary shear stress applied to the wetted perimeter is equal to:

$$\tau_o = \gamma \, RS_o \qquad\qquad (2.3)$$

where,

τ_o = mean boundary shear stress, N/m^2 (lb/ft^2)

γ = unit weight of water, 9810 N/m^3 (62.4 lb/ft^3)

R = hydraulic radius, m (ft)

S_o = average bottom slope (equal to energy slope for uniform flow), m/m (ft/ft)

Shear stress in channels is not uniformly distributed along the wetted perimeter (USBR, 1951; Olsen and Florey, 1952; Chow, 1959; Anderson, et al., 1970). A typical distribution of shear stress in a prismatic channel is shown in Figure 2.1. The shear stress is zero at the water surface and reaches a maximum on the centerline of the channel. The maximum for the side slopes occurs at about the lower third of the side.

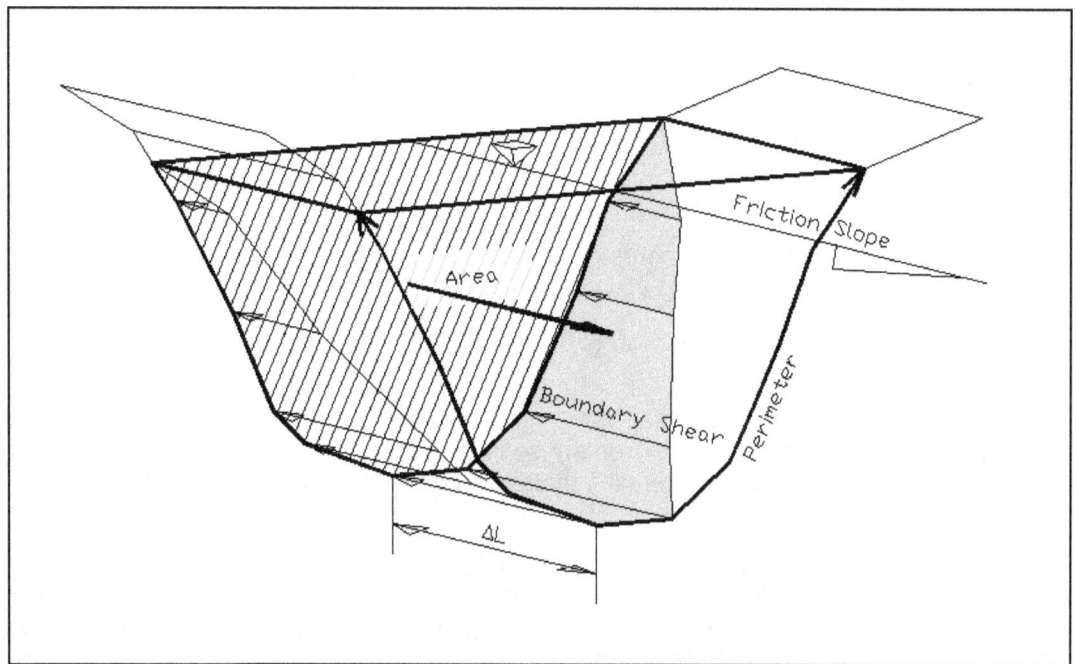

Figure 2.1. Typical Distribution of Shear Stress

The maximum shear stress on a channel bottom, τ_d, and on the channel side, τ_s, in a straight channel depends on the channel shape. To simplify the design process, the maximum channel bottom shear stress is taken as:

$$\tau_d = \gamma\, dS_o \tag{2.4}$$

where,

τ_d = shear stress in channel at maximum depth, N/m^2 (lb/ft^2)

d = maximum depth of flow in the channel for the design discharge, m (ft)

For trapezoidal channels where the ratio of bottom width to flow depth (B/d) is greater than 4, Equation 2.4 provides an appropriate design value for shear stress on a channel bottom. Most roadside channels are characterized by this relatively shallow flow compared to channel width. For trapezoidal channels with a B/d ratio less than 4, Equation 2.4 is conservative. For example, for a B/d ratio of 3, Equation 2.4 overestimates actual bottom shear stress by 3 to 5 percent for side slope values (Z) of 6 to 1.5, respectively. For a B/d ratio of 1, Equation 2.5 overestimates actual bottom shear stress by 24 to 35 percent for the same side slope values of 6 to 1.5, respectively. In general, Equation 2.4 overestimates in cases of relatively narrow channels with steep side slopes.

The relationship between permissible shear stress and permissible velocity for a lining can be found by considering the continuity equation:

$$Q = VA \qquad (2.5)$$

where,

  V = flow velocity, m/s (ft/s)

  A = area of flow, m^2 (ft^2)

By substituting Equation 2.4 and Equation 2.5 into Equation 2.1:

$$V_p = \frac{\alpha}{n\sqrt{\gamma d}} R^{1/6} \tau_p^{1/2} \qquad (2.6)$$

where,

  V_p = permissible velocity, m/s (ft/s)

  τ_p = permissible shear stress, N/m^2 (lb/ft^2)

  α = unit conversion constant, 1.0 (SI), 1.49 (CU)

It can be seen from this equation that permissible velocity varies with the hydraulic radius. However, permissible velocity is not extremely sensitive to hydraulic radius since the exponent is only 1/6. Furthermore, n will change with hydraulic conditions causing an additional variation in permissible velocity.

The tractive force method has a couple of advantages compared to the permissible velocity method. First, the failure criteria for a particular lining are represented by a single permissible shear stress value that is applicable over a wide range of channel slopes and channel shapes. Second, shear stresses are easily calculated using Equation 2.4. Equation 2.4 is also useful in judging the field performance of a channel lining, because depth and gradient may be easier to measure in the field than channel velocity. The advantage of the permissible velocity approach is that most designers are familiar with velocity ranges and have a "feel" for acceptable conditions.

2.2.3 Permissible Shear Stress

Flexible linings act to reduce the shear stress on the underlying soil surface. For example, a long-term lining of vegetation in good condition can reduce the shear stress on the soil surface by over 90 percent. Transitional linings act in a similar manner as vegetative linings to reduce shear stress. Performance of these products depends on their properties: thickness, cover density, and stiffness.

The erodibility of the underlying soil therefore is a key factor in the performance of flexible linings. The erodibility of soils is a function of particle size, cohesive strength and soil density. The erodibility of non-cohesive soils (defined as soils with a plasticity index of less than 10) is due mainly to particle size, while fine-grained cohesive soils are controlled mainly by cohesive strength and soil density. For most highway construction, the density of the roadway embankment is controlled by compaction rather than the natural density of the undisturbed ground. However, when the ditch is lined with topsoil, the placed density of the topsoil should be used instead of the density of the compacted embankment soil.

For stone linings, the permissible shear stress, τ_p, indicates the force required to initiate movement of the stone particles. Prior to movement of stones, the underlying soil is relatively protected. Therefore permissible shear stress is not significantly affected by the erodibility of the underlying soil. However, if the lining moves, the underlying soil will be exposed to the erosive force of the flow.

Table 2.3 provides typical examples of permissible shear stress for selected lining types. Representative values for different soil types are based on the methods found in Chapter 4 while those for gravel mulch and riprap are based on methods found in Chapter 7. Vegetative and RECP lining performance relates to how well they protect the underlying soil from shear stresses so these linings do not have permissible shear stresses independent of soil types. Chapters 4 (vegetation) and 5 (RECPs) describe the methods for analyzing these linings. Permissible shear stress for gabion mattresses depends on rock size and mattress thickness as is described in Section 7.2.

Table 2.3. Typical Permissible Shear Stresses for Bare Soil and Stone Linings

Lining Category	Lining Type	Permissible Shear Stress	
		N/m^2	lb/ft^2
Bare Soil[1] Cohesive (PI = 10)	Clayey sands	1.8-4.5	0.037-0.095
	Inorganic silts	1.1-4.0	0.027-0.11
	Silty sands	1.1-3.4	0.024-0.072
Bare Soil[1] Cohesive (PI $\geq$ 20)	Clayey sands	4.5	0.094
	Inorganic silts	4.0	0.083
	Silty sands	3.5	0.072
	Inorganic clays	6.6	0.14
Bare Soil[2] Non-cohesive (PI < 10)	Finer than coarse sand D_{75}<1.3 mm (0.05 in)	1.0	0.02
	Fine gravel D_{75}=7.5 mm (0.3 in)	5.6	0.12
	Gravel D_{75}=15 mm (0.6 in)	11	0.24
Gravel Mulch[3]	Coarse gravel D_{50} = 25 mm (1 in)	19	0.4
	Very coarse gravel D_{50} = 50 mm (2 in)	38	0.8
Rock Riprap[3]	D_{50} = 0.15 m (0.5 ft)	113	2.4
	D_{50} = 0.30 m (1.0 ft)	227	4.8

[1]Based on Equation 4.6 assuming a soil void ratio of 0.5 (USDA, 1987).
[2]Based on Equation 4.5 derived from USDA (1987)
[3]Based on Equation 6.7 with Shield's parameter equal to 0.047.

2.3   DESIGN PARAMETERS

2.3.1   Design Discharge Frequency

Design flow rates for permanent roadside and median drainage channel linings usually have a 5 or 10-year return period. A lower return period flow is allowable if a transitional lining is to be used, typically the mean annual storm (approximately a 2-year return period, i.e., 50 percent probability of occurrence in a year). Transitional channel linings are often used during the establishment of vegetation. The probability of damage during this relatively short time is low, and if the lining is damaged, repairs are easily made. Design procedures for determining the maximum permissible discharge in a roadway channel are given in Chapter 3.

2.3.2 Channel Cross Section Geometry

Most highway drainage channels are trapezoidal or triangular in shape with rounded corners. For design purposes, a trapezoidal or triangular representation is sufficient. Design of roadside channels should be integrated with the highway geometric and pavement design to insure proper consideration of safety and pavement drainage needs. If available channel linings are found to be inadequate for the selected channel geometry, it may be feasible to widen the channel. Either increasing the bottom width or flattening the side slopes can accomplish this. Widening the channel will reduce the flow depth and lower the shear stress on the channel perimeter. The width of channels is limited however to the ratio of top width to depth less than about 20 (Richardson, Simons and Julien, 1990). Very wide channels have a tendency to form smaller more efficient channels within their banks, which increase shear stress above planned design range.

It has been demonstrated that if a riprap-lined channel has 1:3 or flatter side slopes, there is no need to check the banks for erosion (Anderson, et al., 1970). With side slopes steeper than 1:3, a combination of shear stress against the bank and the weight of the lining may cause erosion on the banks before the channel bottom is disturbed. The design method in this manual includes procedures for checking the adequacy of channels with steep side slopes.

Equations for determining cross-sectional area, wetted perimeter, and top width of channel geometries commonly used for highway drainage channels are given in Appendix B.

2.3.3 Channel Slope

The slope of a roadside channel is usually the same as the roadway profile and so is not a design option. If channel stability conditions are below the required performance and available linings are nearly sufficient, it may be feasible to reduce the channel slope slightly relative to the roadway profile. For channels outside the roadway right-of-way, there can be more grading design options to adjust channel slope where necessary.

Channel slope is one of the major parameters in determining shear stress. For a given design discharge, the shear stress in the channel with a mild or subcritical slope is smaller than a channel with a supercritical slope. Roadside channels with gradients in excess of about two percent will usually flow in a supercritical state.

2.3.4 Freeboard

The freeboard of a channel is the vertical distance from the water surface to the top of the channel at design condition. The importance of this factor depends on the consequence of overflow of the channel bank. At a minimum, the freeboard should be sufficient to prevent waves or fluctuations in water surface from washing over the sides. In a permanent roadway channel, about 0.15 m (0.5 ft) of freeboard should be adequate, and for transitional channels, zero freeboard may be acceptable. Steep gradient channels should have a freeboard height equal to the flow depth. This allows for large variations to occur in flow depth for steep channels caused by waves, splashing and surging. Lining materials should extend to the freeboard elevation.

CHAPTER 3: GENERAL DESIGN PROCEDURE

This chapter outlines the general design procedure for flexible channel linings based on design concepts presented in Chapter 2. The simplest case of the straight channel is described first. Subsequent sections consider variations to the straight channel including side slope stability, composite linings, and bends. The final two sections address additional considerations for channels with a steep longitudinal slope and determination of a maximum discharge for a given channel. This chapter is intended to apply to all flexible linings. Subsequent chapters provide more detailed guidance on specific flexible lining types.

3.1 STRAIGHT CHANNELS

The basic design procedure for flexible channel linings is quite simple. The computations include a determination of the uniform flow depth in the channel, known as the normal depth, and determination of the shear stress on the channel bottom at this depth. Both concepts were discussed in Chapter 2. Recalling Equation 2.8, the maximum shear stress is given by:

$$\tau_d = \gamma d S_o \qquad (3.1)$$

where,

τ_d = shear stress in channel at maximum depth, N/m^2 (lb/ft^2)

γ = unit weight of water, N/m^3 (lb/ft^3)

d = depth of flow in channel, m (ft)

S_o = channel bottom slope, m/m (ft/ft)

The basic comparison required in the design procedure is that of permissible to computed shear stress for a lining. If the permissible shear stress is greater than or equal to the computed shear stress, including consideration of a safety factor, the lining is considered acceptable. If a lining is unacceptable, a lining with a higher permissible shear stress is selected, the discharge is reduced (by diversion or retention/detention), or the channel geometry is modified. This concept is expressed as:

$$\tau_p \geq SF \; \tau_d \qquad (3.2)$$

where,

τ_p = permissible shear stress for the channel lining, N/m^2 (lb/ft^2)

SF = safety factor (greater than or equal to one)

τ_d = shear stress in channel at maximum depth, N/m^2 (lb/ft^2)

The safety factor provides for a measure of uncertainty, as well as a means for the designer to reflect a lower tolerance for failure by choosing a higher safety factor. A safety factor of 1.0 is appropriate in many cases and may be considered the default. The expression for shear stress at maximum depth (Equation 3.1) is conservative and appropriate for design as discussed in Chapter 2. However, safety factors from 1.0 to 1.5 may be appropriate, subject to the designer's discretion, where one or more of the following conditions may exist:

- critical or supercritical flows are expected

- climatic regions where vegetation may be uneven or slow to establish

- significant uncertainty regarding the design discharge

- consequences of failure are high

The basic procedure for flexible lining design consists of the following steps and is summarized in Figure 3.1. (An alternative process for determining an allowable discharge given slope and shape is presented in Section 3.6.)

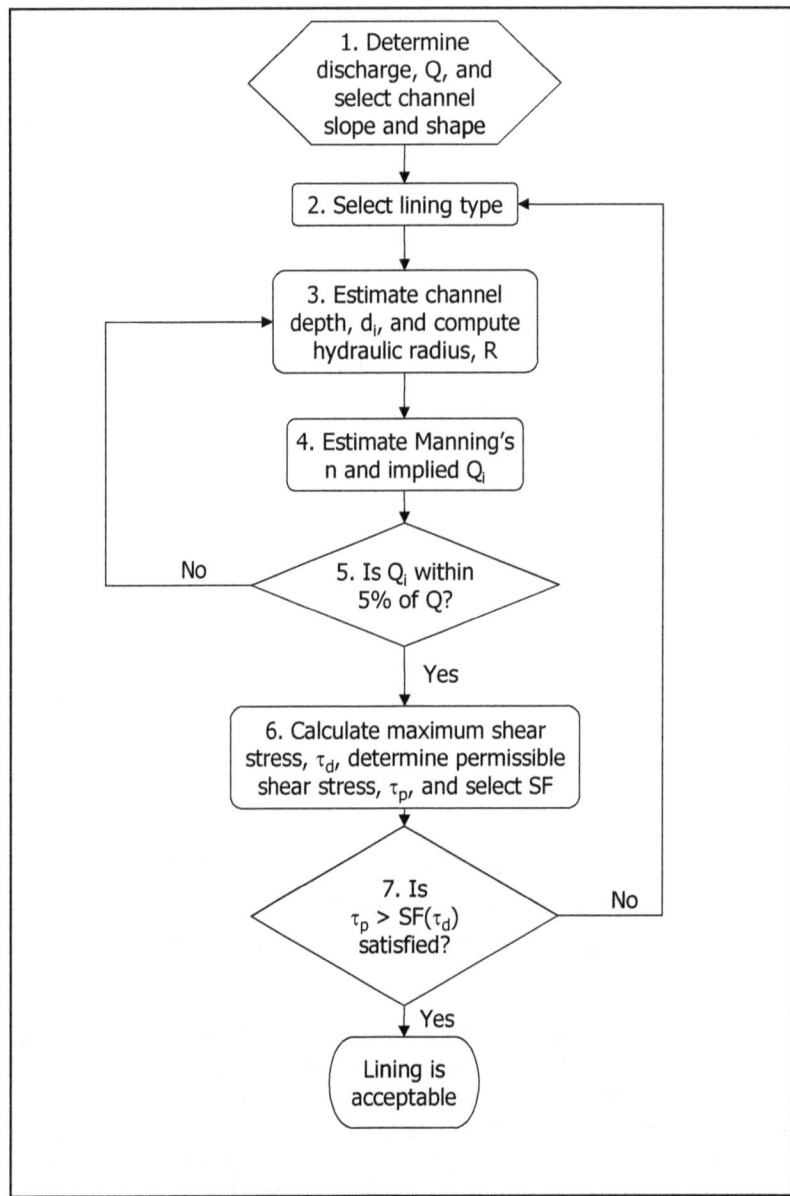

Figure 3.1. Flexible Channel Lining Design Flow Chart

Step 1. Determine a design discharge and select the channel slope and channel shape.

Step 2. Select a trial lining type. Initially, the designer may need to determine if a long-term lining is needed and whether or not a temporary or transitional lining is required. For determining the latter, the trial lining type could be chosen as the native material (unlined), typically bare soil. For example, it may be determined that the bare soil is insufficient for a long-term solution, but vegetation is a good solution. For the transitional period between construction and vegetative establishment, analysis of the bare soil will determine if a temporary lining is prudent.

Step 3. Estimate the depth of flow, d_i in the channel and compute the hydraulic radius, R. The estimated depth may be based on physical limits of the channel, but this first estimate is essentially a guess. Iterations on steps 3 through 5 may be required.

Step 4. Estimate Manning's n and the discharge implied by the estimated n and flow depth values. See Chapters 4 through 7 depending on lining type of interest, Table 2.1, or Table 2.2 for Manning's n values. Calculate the discharge, Q_i.

Step 5. Compare Q_i with Q. If Q_i is within 5 percent of the design Q then proceed on to Step 6. If not, return to step 3 and select a new estimated flow depth, d_{i+1}. This can be estimated from the following equation or any other appropriate method.

$$d_{i+1} = d_i \left(\frac{Q}{Q_i} \right)^{0.4}$$

Step 6. Calculate the shear stress at maximum depth, τ_d (Equation 3.1), determine the permissible shear stress, τ_p, and select an appropriate safety factor. Permissible shear stress is determined based on guidance in Chapters 4 through 7, as applicable to the chosen lining, or Table 2.3. A safety factor of 1.0 is usually chosen, but may be increased as discussed earlier.

Step 7. Compare the permissible shear stress to the calculated shear stress from step 6 using Equation 3.2. If the permissible shear stress is adequate then the lining is acceptable. If the permissible shear is inadequate, then return to step 2 and select an alternative lining type with greater permissible shear stress. As an alternative, a different channel shape may be selected that results in a lower depth of flow.

The selected lining is stable and the design process is complete. Other linings may be tested, if desired, before specifying the preferred lining.

Design Example: Basic Channel (SI)

Evaluate a proposed gravel mulch lining on a trapezoidal channel for stability. Given:

Q   = 0.42 m³/s

B   = 0.4 m

Z   = 3

S_o = 0.008 m/m

D_{50} = 25 mm

Solution

Step 1. Channel slope, shape and discharge have been given.

Step 2. Proposed lining type is a gravel mulch with D_{50} = 25 mm.

Step 3. Assume that the depth of flow, d_i in the channel is 0.5 m. Compute R. The equations in Appendix B may be used for this.

$$A = Bd + Zd^2 = 0.4(0.5) + 3(0.5)^2 = 0.950 \text{ m}^2$$

$$P = B + 2d\sqrt{Z^2 + 1} = 0.4 + 2(0.5)\sqrt{3^2 + 1} = 3.56 \text{ m}$$

$$R = A/P = 0.950/3.56 = 0.267 \text{ m}$$

Step 4. From Table 2.2, Manning's n equals 0.033. (Equations 6.1 or 6.2 should be used for this specific site, but for ease of illustration the value from Table 2.2 is used in this example.) The discharge is calculated using Manning's equation (Equation 2.1):

$$Q = \frac{\alpha}{n} A R^{\frac{2}{3}} S_f^{\frac{1}{2}} = \frac{1}{0.033}(0.950)(0.267)^{\frac{2}{3}}(0.008)^{\frac{1}{2}} = 1.07 \text{ m}^3 / s$$

Step 5. Since this value is more than 5 percent different from the design flow, we need to go back to step 3 to estimate a new flow depth.

Step 3 (2nd iteration). Estimate a new depth estimate:

$$d_{i+1} = d_i \left(\frac{Q}{Q_i}\right)^{0.4} = 0.500\left(\frac{0.42}{1.07}\right)^{0.4} = 0.344 \text{ m}$$

Compute new hydraulic radius.

$$A = Bd + Zd^2 = 0.4(0.344) + 3(0.344)^2 = 0.493 \text{ m}^2$$

$$P = B + 2d\sqrt{Z^2 + 1} = 0.4 + 2(0.344)\sqrt{3^2 + 1} = 2.58 \text{ m}$$

$$R = A/P = 0.493/2.58 = 0.191 \text{ m}$$

Step 4 (2nd iteration). Table 2.2 does not have a 0.344 m depth so Equation 6.1 is used for estimating Manning's n. Manning's n equals 0.035. The discharge is calculated using Manning's equation:

$$Q = \frac{\alpha}{n} A R^{\frac{2}{3}} S_f^{\frac{1}{2}} = \frac{1}{0.035}(0.493)(0.191)^{\frac{2}{3}}(0.008)^{\frac{1}{2}} = 0.42 \text{ m}^3 / s$$

Step 5 (2nd iteration). Since this value is within 5 percent of the design flow (we hit it right on), we can now proceed to step 6.

Step 6. The shear stress at maximum depth from Equation 3.1 is:

$$\tau_d = \gamma d S_o = 9810(0.344)(0.008) = 27 \text{ N/m}^2$$

From Table 2.3, the permissible shear stress, τ_p = 19 N/m^2.

For this channel, a SF = 1.0 is chosen.

Step 7. Compare calculated shear to permissible shear using Equation 3.2:

$$\tau_p \geq SF \, \tau_d$$

$19 \geq 1.0$ (27) No. The lining is not stable! Go back to step 2. Select an alternative lining type with greater permissible shear stress. Try the next larger size of gravel. If the lining had been stable, the design process would be complete.

Design Example: Basic Channel (CU)

Evaluate a proposed gravel mulch lining on a trapezoidal channel for stability. Given:

Q	=	15 ft^3/s
B	=	1.3 ft
Z	=	3
S$_o$	=	0.008 ft/ft
D$_{50}$	=	1 in

Solution

Step 1. Channel slope, shape and discharge have been given.

Step 2. Proposed lining type is a gravel mulch with $D_{50} = 1$ in.

Step 3. Assume that the depth of flow, d_i in the channel is 1.6 ft. Compute R. The equations in Appendix B may be used for this.

$$A = Bd+Zd^2 = 1.3(1.6)+3(1.6)^2 = 9.76 \text{ ft}^2$$

$$P = B + 2d\sqrt{Z^2 +1} = 1.3 + 2(1.6)\sqrt{3^2 +1} = 11.4 \text{ ft}$$

$$R = A/P = 9.76/11.4 = 0.856 \text{ ft}$$

Step 4. From Table 2.2, Manning's n equals 0.033. (Equations 6.1 or 6.2 should be used for this specific site, but for ease of illustration the value from Table 2.2 is used in this example.) The discharge is calculated using Manning's equation (Equation 2.1):

$$Q = \frac{\alpha}{n} AR^{\frac{2}{3}}S_f^{\frac{1}{2}} = \frac{1.49}{0.033}(9.76)(0.856)^{\frac{2}{3}}(0.008)^{\frac{1}{2}} = 35.5 \text{ ft}^3/s$$

Step 5. Since this value is more than 5 percent different from the design flow, we need to go back to step 3 to estimate a new flow depth.

Step 3 (2nd iteration). Estimate a new depth estimate:

$$d_{i+1} = d_i \left(\frac{Q}{Q_i}\right)^{0.4} = 1.6\left(\frac{15}{35.5}\right)^{0.4} = 1.13 \text{ ft}$$

Compute a new hydraulic radius.

$$A = Bd+Zd^2 = 1.3(1.13)+3(1.13)^2 = 5.30 \text{ ft}^2$$

$$P = B + 2d\sqrt{Z^2 +1} = 1.3 + 2(1.13)\sqrt{3^2 +1} = 8.45 \text{ ft}$$

$$R = A/P = 5.30/8.45 = 0.627 \text{ ft}$$

Step 4 (2nd iteration). Table 2.2 does not have a 1.13 ft depth so Equation 6.1 is used for estimating Manning's n. Manning's n equals 0.035. The discharge is calculated using Manning's equation:

$$Q = \frac{\alpha}{n} AR^{\frac{2}{3}} S_f^{\frac{1}{2}} = \frac{1.49}{0.035}(5.3)(0.627)^{\frac{2}{3}}(0.008)^{\frac{1}{2}} = 14.8 \text{ ft}^3 / s$$

Step 5. (2nd iteration). Since this value is within 5 percent of the design flow, we can now proceed to step 6.

Step 6. The shear stress at maximum depth from Equation 3.1 is:

$\tau_d = \gamma dS_o = 62.4(1.13)(0.008) = 0.56 \text{ lb/ft}^2$

From Table 2.3, permissible shear stress, $\tau_p = 0.4 \text{ lb/ft}^2$.

For this channel, a SF = 1.0 is chosen.

Step 7. Compare calculated shear to permissible shear using Equation 3.2:

$\tau_p \geq SF \tau_d$

$0.40 \geq 1.0 (0.56)$ No. The lining is not stable! Go back to step 2. Select an alternative lining type with greater permissible shear stress. Try the next larger size of gravel. If the lining had been stable, the design process would be complete.

3.2 SIDE SLOPE STABILITY

As described in Chapter 2, shear stress is generally reduced on the channel sides compared with the channel bottom. The maximum shear on the side of a channel is given by the following equation:

$$\tau_s = K_1 \tau_d \tag{3.3}$$

where,

τ_s = side shear stress on the channel, N/m^2 (lb/ft^2)

K_1 = ratio of channel side to bottom shear stress

τ_d = shear stress in channel at maximum depth, N/m^2 (lb/ft^2)

The value K_1 depends on the size and shape of the channel. For parabolic or V-shape with rounded bottom channels there is no sharp discontinuity along the wetted perimeter and therefore it can be assumed that shear stress at any point on the side slope is related to the depth at that point using Equation 3.1.

For trapezoidal and triangular channels, K_1 has been developed based on the work of Anderson, et al. (1970). The following equation may be applied.

$$\begin{array}{lll} K_1 = 0.77 & Z \leq 1.5 & \\ K_1 = 0.066Z + 0.67 & 1.5 < Z < 5 & (3.4) \\ K_1 = 1.0 & 5 \leq Z & \end{array}$$

The Z value represents the horizontal dimension 1:Z (V:H). Use of side slopes steeper than 1:3 (V:H) is not encouraged for flexible linings other than riprap or gabions because of the potential for erosion of the side slopes. Steep side slopes are allowable within a channel if cohesive soil conditions exist. Channels with steep slopes should not be allowed if the channel is constructed in non-cohesive soils.

For riprap and gabions, the basic design procedure is supplemented for channels with side slopes steeper than 1:3 in Chapters 6 and 7, respectively.

3.3 COMPOSITE LINING DESIGN

Composite linings use two lining types in a single channel rather than one. A more shear resistant lining is used in the bottom of the channel while a less shear resistant lining protects the sides. This type of design may be desirable where the upper lining is more cost-effective and/or environmentally benign, but the lower lining is needed to resist bottom stresses.

Another important use of a composite lining is in vegetative channels that experience frequent low flows. These low flows may kill the submerged vegetation. In erodible soils, this leads to the formation of a small gully at the bottom of the channel. Gullies weaken a vegetative lining during higher flows, causing additional erosion, and can result in a safety hazard. A solution is to provide a non-vegetative low-flow channel lining such as concrete or riprap. The dimensions of the low-flow channel are sufficient to carry frequent low flows but only a small portion of the design flow. The remainder of the channel is covered with vegetation.

It is important that the bottom lining material cover the entire channel bottom so that adequate protection is provided. To insure that the channel bottom is completely protected, the bottom lining should be extended a small distance up the side slope.

Computation of flow conditions in a composite channel requires the use of an equivalent Manning's n value for the entire perimeter of the channel. For determination of equivalent roughness, the channel area is divided into two parts of which the wetted perimeters and Manning's n values of the low-flow section and channel sides are known. These two areas of the channel are then assumed to have the same mean velocity. The following equation is used to determine the equivalent roughness coefficient, n_e.

$$n_e = \left[\frac{P_L}{P} + \left(1 - \frac{P_L}{P}\right)\left(\frac{n_s}{n_L}\right)^{3/2} \right]^{2/3} n_L \qquad (3.5)$$

where,

n_e = effective Manning's n value for the composite channel
P_L = low flow lining perimeter, m (ft)
P = total flow perimeter, m (ft)
n_s = Manning's n value for the side slope lining
n_L = Manning's n value for the low flow lining

When two lining materials with significantly different roughness values are adjacent to each other, erosion may occur near the boundary of the two linings. Erosion of the weaker lining material may damage the lining as a whole. In the case of composite channel linings with vegetation on the banks, this problem can occur in the early stages of vegetative establishment. A transitional lining should be used adjacent to the low-flow channel to provide erosion protection until the vegetative lining is well established.

The procedure for composite lining design is based on the design procedure presented in Section 3.1 with additional sub-steps to account for the two lining types. Specifically, the modifications are:

Step 1. Determine design discharge and select channel slope and shape. (No change.)

Step 2. Need to select both a low flow and side slope lining.

Step 3. Estimate the depth of flow in the channel and compute the hydraulic radius. (No change.)

Step 4. After determining the Manning's n for the low flow and side slope linings, use Equation 3.5 to calculate the effective Manning's n.

Step 5. Compare implied discharge and design discharge. (No change.)

Step 6. Determine the shear stress at maximum depth, τ_d (Equation 3.1), and the shear stress on the channel side slope, τ_s (Equation 3.3).

Step 7. Compare the shear stresses, τ_d and τ_s, to the permissible shear stress, τ_p, for each of the channel linings. If τ_d or τ_s is greater than the τ_p for the respective lining, a different combination of linings should be evaluated.

Design Example: Composite Lining Design (SI)

Evaluate the channel design for the composite concrete and vegetation lining given in Figure 3.2. Given:

Q = 0.28 m³/s
B = 0.9 m   Concrete low flow channel
Z = 3
S_o = 0.02 m/m
Vegetation: Class C, height = 0.2 m (mixed with good cover)

The underlying soil is a clayey sand (SC) soil with a plasticity index of 16 and a porosity of 0.5. (See Section 4.3.2.)

Solution

Step 1. Channel slope, shape and discharge have been given.

Step 2. Low flow lining is concrete. Side slope lining is Class C vegetation.

Step 3. Assume that the depth of flow, d_i in the channel is 0.30 m. Determine R. Assume that the concrete portion is essentially flat.

$$A = Bd+Zd^2 = 0.9(0.3)+3(0.3)^2 = 0.540 \text{ m}^2$$

$$P = B + 2d\sqrt{Z^2 +1} = 0.9 + 2(0.3)\sqrt{3^2 +1} = 2.80 \text{ m}$$

R = A/P = 0.540/2.80 = 0.193 m

Step 4. From the methods in Chapter 4 (Table 4.4 and Equation 4.2) the n value for Class C vegetation is determined to be n_s = 0.043. For concrete, the typical n value for concrete from Table 2.1 is n_L = 0.013.

Effective Manning's n is calculated using Equation 3.5.

$$n_e = \left[\frac{P_L}{P}+\left(1-\frac{P_L}{P}\right)\left(\frac{n_s}{n_L}\right)^{3/2}\right]^{2/3} n_L = \left[\frac{0.9}{2.80}+\left(1-\frac{0.9}{2.80}\right)\left(\frac{0.043}{0.013}\right)^{3/2}\right]^{2/3} 0.013 = 0.035$$

Calculate flow using Manning's equation (Equation 2.1):

$$Q = \frac{\alpha}{n}(A)(R)^{\frac{2}{3}}(S)^{\frac{1}{2}} = \frac{1}{0.035}(0.540)(0.193)^{\frac{2}{3}}(0.02)^{\frac{1}{2}} = 0.73 \text{ m}^3/\text{s}$$

Step 5. This flow is larger than the design flow of 0.28 m³/s by more than 5 percent. Go back to step 3.

Step 3 (2nd iteration). Estimate a new depth from an appropriate method. Try d=0.19 m.

Determine R with the new depth.

$A = Bd+Zd^2 = 0.9(0.19)+3(0.19)^2 = 0.279 \text{ m}^2$

$P = B + 2d\sqrt{Z^2 + 1} = 0.9 + 2(0.19)\sqrt{3^2 + 1} = 2.10 \text{ m}$

R = A/P = 0.279/2.10 = 0.133 m

Step 4 (2nd iteration). From the methods in Chapter 4 (Table 4.4 and Equation 4.2) the n value for Class C vegetation is determined to be n_s = 0.052. For concrete, the n value is the same n_L = 0.013.

Effective Manning's n is calculated using Equation 3.5.

$$n_e = \left[\frac{P_L}{P} + \left(1 - \frac{P_L}{P}\right)\left(\frac{n_S}{n_L}\right)^{\frac{3}{2}}\right]^{\frac{2}{3}} n_L = \left[\frac{0.9}{2.10} + \left(1 - \frac{0.9}{2.10}\right)\left(\frac{0.052}{0.013}\right)^{\frac{3}{2}}\right]^{\frac{2}{3}} 0.013 = 0.038$$

Calculate flow using Manning's equation:

$$Q = \frac{1}{0.038}(0.279)(0.133)^{\frac{2}{3}}(0.02)^{\frac{1}{2}} = 0.27 \text{ m}^3/\text{s}$$

Step 5 (2nd iteration). This flow is within 5 % of the design discharge, therefore, proceed to step 6.

Step 6. Calculate maximum shear stress, determine permissible shear stress, and select SF.

Bottom shear stress is calculated from Equation 3.1:

$$\tau_d = \gamma d S_o = 9810(0.19)(0.02) = 37 \text{ N}/\text{m}^2$$

Maximum side shear stress is calculated from Equation 3.3 after calculating K_1:

$$K_1 = 0.066Z + 0.67 = 0.066(3) + 0.67 = 0.87$$

$$\tau_s = K_1\tau_d = 0.87(37) = 32 \text{ N}/\text{m}^2$$

Concrete is a non-erodible, rigid lining so it has a very high rigid permissible shear stress. By using the techniques in Section 4.3 (Equation 4.7 and Table 4.5) the permissible shear stress of the vegetative portion of the lining is 123 N/m².

A safety factor of 1.0 is chosen for this situation.

Step 7. The maximum shear stress on the channel side slopes (32 N/m²) is less than permissible shear stress on the vegetation (123 N/m²) so the lining is acceptable. (The concrete bottom lining is non-erodible.)

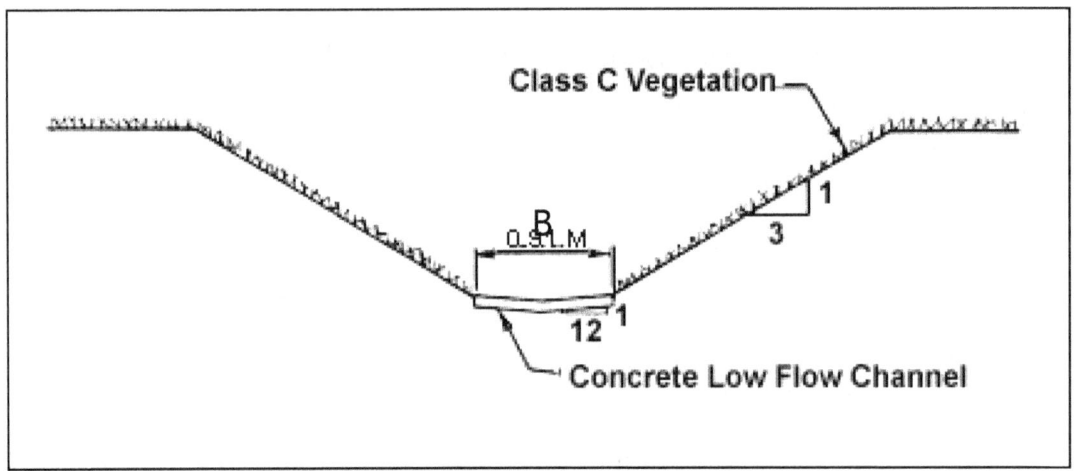

Figure 3.2. Composite Lining Design Example

Design Example: Composite Lining Design (CU)

Evaluate the channel design for the composite concrete and vegetation lining given in Figure 3.2. Given:

Q $\quad$ = 9.9 ft^3/s

B $\quad$ = 3 ft $\quad$ Concrete low flow channel

Z $\quad$ = 3

S_o $\quad$ = 0.02 ft/ft

Vegetation: Class C, height = 0.66 ft (mixed with good cover)

The underlying soil is a clayey sand (SC) soil with a plasticity index of 16 and a porosity of 0.5 (See Section 4.3.2)

Solution

Step 1. Channel slope, shape and discharge have been given.

Step 2. Low flow lining is concrete. Side slope lining is Class C vegetation.

Step 3. Assume that the depth of flow, d_i in the channel is 1.0 ft. Determine R. Assume that the concrete portion is essentially flat.

$A = Bd+Zd^2 = 3.0(1.0)+3(1.0)^2 = 6.00$ ft^2

$P = B + 2d\sqrt{Z^2 + 1} = 3.0 + 2(1.0)\sqrt{3^2 + 1} = 9.32$ ft

$R = A/P = 6.00/9.32 = 0.643$ ft

Step 4. From the methods in Chapter 4 (Table 4.4 and Equation 4.2) the n value for Class C vegetation is determined to be $n_s = 0.043$. For concrete, the typical n value for concrete from Table 2.1 is $n_L = 0.013$.

Effective Manning's n is calculated using Equation 3.5.

$$n_e = \left[\frac{P_L}{P} + \left(1 - \frac{P_L}{P}\right)\left(\frac{n_S}{n_L}\right)^{3/2}\right]^{2/3} n_L = \left[\frac{3.0}{9.32} + \left(1 - \frac{3.0}{9.32}\right)\left(\frac{0.043}{0.013}\right)^{3/2}\right]^{2/3} 0.013 = 0.035$$

Calculate flow using Manning's equation (Equation 2.1):

$$Q = \frac{\alpha}{n}(A)(R)^{2/3}(S)^{1/2} = \frac{1.49}{0.035}(6.00)(0.643)^{2/3}(0.02)^{1/2} = 26.9 \text{ ft}^3 / s$$

Step 5.   This flow is larger than the design flow of 9.9 ft³/s by more than 5 percent. Go back to step 3.

Step 3 (2nd iteration). Estimate a new depth from an appropriate approach. Try d = 0.62 ft.

Determine R with the new depth.

$A = Bd + Zd^2 = 3.0(0.62) + 3(0.62)^2 = 3.01 \text{ ft}^2$

$P = B + 2d\sqrt{Z^2 + 1} = 3.0 + 2(0.62)\sqrt{3^2 + 1} = 6.92 \text{ ft}$

R = A/P = 3.01/6.92 = 0.435 ft

Step 4 (2nd iteration).   From the methods in Chapter 4 (Table 4.4 and Equation 4.2) the n value for Class C vegetation is determined to be n_s = 0.052. For concrete, the n value is the same n_L = 0.013.

Effective Manning's n is calculated using Equation 3.5.

$$n_e = \left[\frac{P_L}{P} + \left(1 - \frac{P_L}{P}\right)\left(\frac{n_S}{n_L}\right)^{3/2}\right]^{2/3} n_L = \left[\frac{3.0}{6.92} + \left(1 - \frac{3.0}{6.92}\right)\left(\frac{0.052}{0.013}\right)^{3/2}\right]^{2/3} 0.013 = 0.038$$

Calculate flow using Manning's equation:

$$Q = \frac{1.49}{0.038}(3.01)(0.435)^{2/3}(0.02)^{1/2} = 9.6 \text{ ft}^3 / s$$

Step 5 (2nd iteration). This flow is within 5% of the design discharge, therefore, proceed to step 6.

Step 6.   Calculate maximum shear stress, determine permissible shear stress, and select SF.

Bottom shear stress is calculated from Equation 3.1:

$\tau_d = \gamma d S_o = 62.4(0.62)(0.02) = 0.77 \text{ lb} / \text{ft}^2$

Maximum side shear stress is calculated from Equation 3.3 after calculating K_1:

$K_1 = 0.066Z + 0.67 = 0.066(3) + 0.67 = 0.87$

$\tau_s = K_1\tau_d = 0.87(0.77) = 0.67 \text{ lb} / \text{ft}^2$

Concrete is a non-erodible, rigid lining so it has a very high rigid permissible shear stress. By using the techniques in Section 4.3 (Equation 4.7 and Table 4.5) the permissible shear stress of the vegetative portion of the lining is 2.5 lb/ft^2.

A safety factor of 1.0 is chosen for this situation.

Step 7. The maximum shear stress on the channel side slopes (0.67 lb/ft^2) is less than permissible shear stress on the vegetation (2.5 lb/ft^2) so the lining is acceptable. (The concrete bottom lining is non-erodible.)

3.4 STABILITY IN BENDS

Flow around a bend creates secondary currents, which impose higher shear stresses on the channel sides and bottom compared to a straight reach (Nouh and Townsend, 1979) as shown in Figure 3.3. At the beginning of the bend, the maximum shear stress is near the inside and moves toward the outside as the flow leaves the bend. The increased shear stress caused by a bend persists downstream of the bend.

Equation 3.6 gives the maximum shear stress in a bend.

$$\tau_b = K_b \tau_d \tag{3.6}$$

where,

τ_b = side shear stress on the channel, N/m^2 (lb/ft^2)

K_b = ratio of channel bend to bottom shear stress

τ_d = shear stress in channel at maximum depth, N/m^2 (lb/ft^2)

The maximum shear stress in a bend is a function of the ratio of channel curvature to the top (water surface) width, R_C/T. As R_C/T decreases, that is as the bend becomes sharper, the maximum shear stress in the bend tends to increase. K_b can be determined from the following equation from Young, et al., (1996) adapted from Lane (1955):

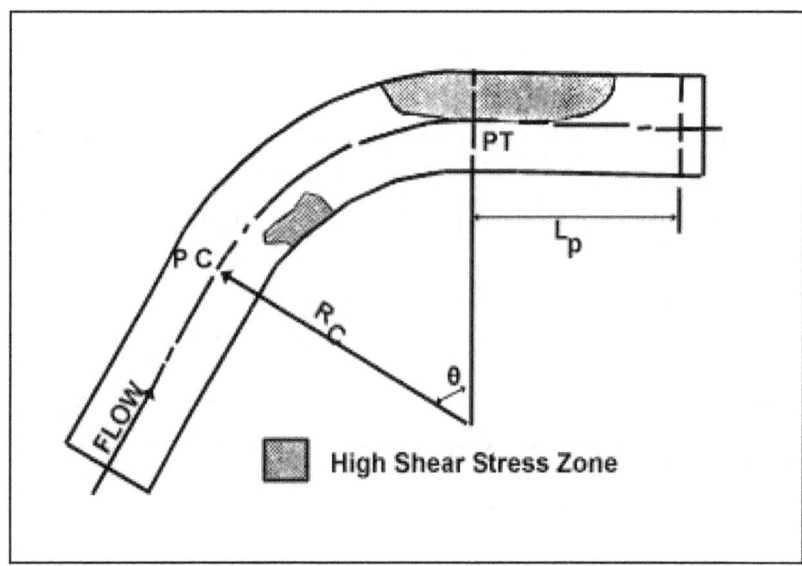

Figure 3.3. Shear Stress Distribution in a Channel Bend (Nouh and Townsend, 1979)

$$K_b = 2.00 \qquad\qquad\qquad\qquad\qquad R_C/T \le 2$$

$$K_b = 2.38 - 0.206\left(\frac{R_c}{T}\right) + 0.0073\left(\frac{R_c}{T}\right)^2 \qquad 2 < R_C/T < 10 \qquad (3.7)$$

$$K_b = 1.05 \qquad\qquad\qquad\qquad\qquad 10 \le R_C/T$$

where,

    R_c = radius of curvature of the bend to the channel centerline, m (ft)

    T   = channel top (water surface) width, m (ft)

The added stress induced by bends does not fully attenuate until some distance downstream of the bend. If added lining protection is needed to resist the bend stresses, this protection should continue downstream a length given by:

$$L_p = \alpha\left(\frac{R^{7/6}}{n}\right) \qquad\qquad (3.8)$$

where,

    L_p = length of protection, m (ft)

    R   = hydraulic radius of the channel, m (ft)

    n   = Manning's roughness for lining material in the bend

    α = unit conversion constant, 0.74 (SI) and 0.60 (CU)

A final consideration for channel design at bends is the increase in water surface elevation at the outside of the bend caused by the superelevation of the water surface. Additional freeboard is necessary in bends and can be calculated use the following equation:

$$\Delta d = \frac{V^2 T}{g R_c} \qquad\qquad (3.9)$$

where,

    Δd = additional freeboard required because of superelevation, m (ft)

    V   = average channel velocity, m/s (ft/s)

    T   = water surface top width, m (ft)

    g   = acceleration due to gravity, m/s^2 (ft/s^2)

    R_c = radius of curvature of the bend to the channel centerline, m (ft)

The design procedure for channel bends is summarized in the following steps:

    Step 1. Determine the shear stress in the bend and check whether or not an alternative lining is needed in the bend.

    Step 2. If an alternative lining is needed, select a trial lining type and compute the new hydraulic properties and bend shear stress.

    Step 3. Estimate the required length of protection.

Step 4. Calculate superelevation and check freeboard in the channel.

Design Example: Channel Bends (SI)

Determine an acceptable channel lining for a trapezoidal roadside channel with a bend. Also compute the necessary length of protection and the superelevation. The location is shown in Figure 3.4. A riprap lining (D_{50} = 0.15 m) has been used on the approaching straight channel (τ_p = 113 N/m^2 from Table 2.3).

Given:

Q	=	0.55 m^3/s
d	=	0.371 m
T	=	3.42 m
B	=	1.2 m
Z	=	3
S$_o$	=	0.015 m/m
R$_C$	=	10 m

Shear stress in the approach straight reach, τ_d = 54.5 N/m^2

Solution

Step 1. Determine the shear stress in the bend using Equation 3.6. First, calculate Kb from Equation 3.7:

$$K_b = 2.38 - 0.206\left(\frac{R_c}{T}\right) + 0.0073\left(\frac{R_c}{T}\right)^2 = 2.38 - 0.206\left(\frac{10.0}{3.42}\right) + 0.0073\left(\frac{10.0}{3.42}\right)^2 = 1.84$$

Bend shear stress is then calculated from Equation 3.6:

$$\tau_b = K_b\tau_d = 1.84(54.5) = 100 \, N/m^2$$

Step 2. Compare the bend shear stress with the permissible shear stress. The permissible shear stress has not been exceeded in the bend. Therefore, the lining in the approach channel can be maintained through the bend. If the permissible shear stress had been exceeded, a more resistant lining would need to be evaluated. A new normal depth would need to be found and then Step 1 repeated.

Step 3. Calculate the required length of protection. Since the same lining is being used in the approach channel and the bend, the length of protection is not relevant in this situation. However, we will calculate it to illustrate the process. Using Equation 3.8 and the channel geometrics, the hydraulic radius, R is 0.24 m and n = 0.074 (using Equation 6.1).

$$L_p = \alpha\left(\frac{R^{7/6}}{n}\right) = 0.74\left(\frac{0.24^{7/6}}{0.074}\right) = 1.9 \, m$$

Step 4. Calculate the superelevation of the water surface. First, top width and cross-sectional area must be computed using the geometric properties,

$$T = B + 2Zd = 1.2 + 2(3)(0.371) = 3.42 \, m$$

$$A = Bd + Zd^2 = 1.2(0.371) + 3(0.371)^2 = 0.86 \, m^2$$

The velocity in the channel found using the continuity equation,

V = Q/A = 0.55/0.86 = 0.64 m/s

Solving Equation 3.9,

$$\Delta d = \frac{V^2 T}{g R_c} = \frac{(0.64)^2 (3.42)}{9.81(10.0)} = 0.014 \text{ m}$$

The freeboard in the channel bend should be at least 0.014 meters to accommodate the super elevation of the water surface.

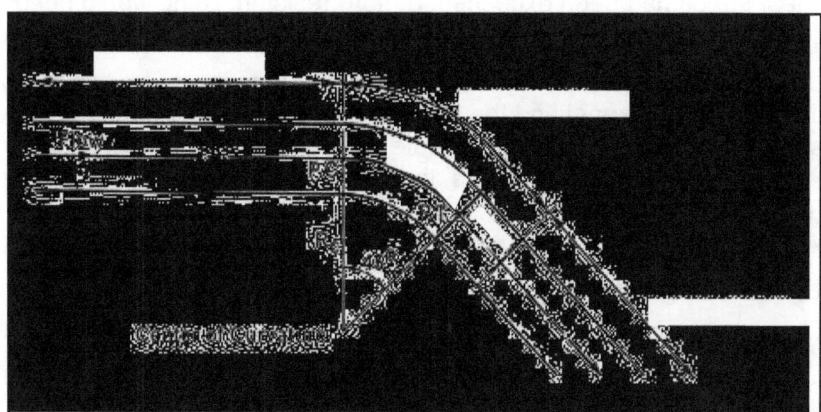

Figure 3.4. Location Sketch of Flexible Linings for Bend Example

Design Example: Channel Bends (CU)

Determine an acceptable channel lining for a trapezoidal roadside channel with a bend. Also compute the necessary length of protection and the superelevation. The location is shown in Figure 3.4. A riprap lining (D_{50} = 0.5 ft) has been used on the approaching straight channel (τ_p = 2.4 lb/ft² from Table 2.3).

Given:

Q	=	20 ft³/s
d	=	1.24 ft
T	=	11.35 ft
B	=	3.9 ft
Z	=	3
S_o	=	0.015 ft/ft
R_C	=	33 ft

Shear stress in the approach straight reach, τ_d = 1.16 lb/ft²

Solution

Step 1. Determine the shear stress in the bend using Equation 3.6. First, calculate K_b from Equation 3.7:

$$K_b = 2.38 - 0.206\left(\frac{R_c}{T}\right) + 0.0073\left(\frac{R_c}{T}\right)^2 = 2.38 - 0.206\left(\frac{33}{11.35}\right) + 0.0073\left(\frac{33}{11.35}\right)^2 = 1.84$$

Bend shear stress is then calculated from Equation 3.6:

$$\tau_b = K_b\tau_d = 1.84(1.16) = 2.13\ \text{lb/ft}^2$$

Step 2. Compare the bend shear stress with the permissible shear stress. The permissible shear stress has not been exceeded in the bend. Therefore, the lining in the approach channel can be maintained through the bend. If the permissible shear stress had been exceeded, a more resistant lining would need to be evaluated. A new normal depth would need to be found and then Step 1 repeated.

Step 3. Calculate the required length of protection. Since the same lining is being used in the approach channel and the bend, the length of protection is not relevant in this situation. However, we will calculate it to illustrate the process. Using Equation 3.8 and the channel geometrics, the hydraulic radius, R is 0.805 ft and n = 0.075 (using Equation 6.1).

$$L_p = \alpha\left(\frac{R^{7\!/\!6}}{n}\right) = 0.60\left(\frac{0.805^{7\!/\!6}}{0.075}\right) = 6.2\ \text{ft}$$

Step 4. Calculate the superelevation of the water surface. First, top width and cross-sectional area must be computed using the geometric properties,

T = B + 2Zd = 3.9 + 2(3)(1.24) = 11.3 ft

A = Bd + Zd² = 3.9(1.24) + 3(1.24)² = 9.45 ft²

The velocity in the channel found using the continuity equation,

V = Q/A = 20/9.45 = 2.12 ft/s

Solving Equation 3.9,

$$\Delta d = \frac{V^2 T}{gR_c} = \frac{(2.12)^2(11.3)}{32.2(33)} = 0.048\ \text{ft}$$

The freeboard in the channel bend should be at least 0.048 ft to accommodate the super elevation of the water surface.

3.5   STEEP SLOPE DESIGN

Intuitively, steep channel slopes may be considered a harsher environment than mild slopes for channel lining design. Furthermore, inspection of Equation 3.1 reveals that applied shear stress is directly proportional to channel slope. Therefore, it is appropriate to address the question of what, if any, additional consideration should be given to flexible channel lining design on steep slopes.

First, "steep" must be defined. From a hydraulic standpoint a steep slope is one that produces a supercritical normal depth (as opposed to a mild slope). Steep may also be defined as a fixed value such as 10 percent. Neither definition is appropriate for all circumstances and a single definition is not required. Two general questions arise when considering steep slopes for channel design.

First, are the same relationships for channel roughness (Manning's n) applicable over the entire range of slopes? For vegetative (Chapter 4) and manufactured (Chapter 5) linings the methodologies for determining roughness do apply to the full range of conditions suitable for these linings; that is, where Equation 3.2 is satisfied. However, for riprap (Chapter 6) and gabion mattress (Chapter 7) linings, slope may influence the approach for estimating roughness. As is discussed in Section 6.1, two roughness relationships are provided: one for relatively shallow flow ($d_a/D_{50} < 1.5$) and one for relatively moderate or deep flow ($d_a/D_{50} > 1.5$). Although slope plays an important role in determining depth; discharge, channel shape, and lining D_{50} also have an influence on the appropriate selection of roughness. (See Section 6.1 for more explanation.)

The second question is whether or not a steeper slope affects the development of the permissible shear stress used in Equation 3.2. Again, for vegetative and manufactured linings, the answer is no and the methods discussed in Chapters 4 and 5, respectively, apply to the full range of conditions where Equation 3.2 is satisfied. The same is also true for gabion linings. In fact, the permissible shear stress relationships presented in Section 7.2 were developed based on testing with slopes up to 33 percent. For riprap linings, Section 6.2 provides two alternative frameworks for evaluating permissible shear stress. One method is for slopes up to 10 percent and the other is for slopes equal to or greater than 10 percent.

Rigid channel linings may be a cost-effective alternative to flexible linings for steep slope conditions. Rigid linings could include asphalt, concrete, or durable bedrock. The decision to select a rigid or flexible lining may be based on other site conditions, such as foundation and maintenance requirements.

For flexible or rigid linings on steep slopes, bends should be avoided. A design requiring a bend in a steep channel should be reevaluated to eliminate the bend, modeled, or designed using an enclosed section.

3.6   MAXIMUM DISCHARGE APPROACH

As the discharge increases along a channel, the shear stress may at some point reach the permissible shear for the channel lining selected indicating the need to proceed with the design of another lining for the next section of channel or provide a relief inlet or culvert to divert the flow out of the channel. The methodology for determining the length or section of channel that a selected lining will remain stable is often referred to as the maximum discharge approach. By knowing the maximum discharge that a lining can sustain, the designer can determine the maximum length of lining for a selected lining type based on the hydrology of the site. This information can assist the designer in an economic evaluation of lining types and can determine relief inlet or culvert spacing.

The procedure presented is for both vegetative linings and non-vegetative linings. Applying the procedure for vegetative linings is particularly useful, since it does not involve a trial and error solution.

Combining Equations 3.1 and 3.2 in the following form can derive the maximum depth a channel lining can withstand:

$$d \leq \frac{\tau_p}{(SF)\gamma S_o} \qquad (3.10)$$

The analysis approach is applied as follows:

Step 1. Select a candidate lining and determine its permissible shear value and an appropriate safety factor.

Step 2. Use Equation 3.10 to calculate the maximum depth. Check that this depth does not exceed the depth (including freeboard) provided in the typical roadway section.

Step 3. Determine the area and hydraulic radius corresponding to the allowable depth based on the channel geometry

Step 4. Estimate the Manning's n value appropriate for the lining type and depth.

Step 5. Solve Manning's equation to determine the maximum discharge for the channel. The length of roadway and/or contributing drainage must be limited to an area generating less than or equal to this amount.

Design Example: Maximum Discharge Approach (SI)

Determine the maximum discharge for a median ditch lined with riprap (D_{50} = 0.150 m). The ditch has a depth of 0.9 m from the roadway shoulder.

Given:

S_o = 0.015 m/m
B   = 1.0 m
Z   = 4
SF = 1.0

Solution

Step 1. Riprap (D_{50} = 0.150 m) has a permissible shear stress of τ_p = 113 N/m² (Table 2.3).

Step 2. Determine the allowable depth from Equation 3.10.

$$d \leq \frac{\tau_p}{(SF)\gamma S_o} = \frac{113}{1.0(9810)(0.015)} = 0.768 \text{ m}$$

The allowable depth is less than the depth of the ditch (0.9 m).

Step 3. Determine the flow area and hydraulic radius from the geometric properties of a trapezoidal channel for the allowable depth:

$$A = Bd + Zd^2 = 1(0.768) + 4(0.768)^2 = 3.13 \text{ m}^2$$

$$P = B + 2d\sqrt{Z^2 + 1} = 1 + 2(0.768)\sqrt{4^2 + 1} = 7.33 \text{ m}$$

R = A/P = 3.13/7.33 = 0.427 m

Step 4. From Equation 6.1, n = 0.059

Step 5. Solving Manning's equation (Equation 2.1):

$$Q = \frac{\alpha}{n}(A)(R)^{\frac{2}{3}}(S)^{\frac{1}{2}} = \frac{1}{0.059}(3.13)(0.427)^{\frac{2}{3}}(0.015)^{\frac{1}{2}} = 3.7 \text{ m}^3/s$$

Flow in this channel must be limited to the calculated flow by providing a relief inlet or culvert or by providing a lining with greater shear resistance.

Design Example: Maximum Discharge Approach (CU)

Determine the maximum discharge for a median ditch lined with riprap (D_{50} = 0.5 ft). The ditch has a depth of 3 ft from the roadway shoulder.

Given:

S_o = 0.015 ft/ft

B = 3.3 ft

Z = 4

SF = 1.0

Solution

Step 1. Riprap (D_{50} = 0.5 ft) has a permissible shear stress of τ_p = 2.4 lb/ft^2 (Table 2.3).

Step 2. Determine the allowable depth from Equation 3.10.

$$d \leq \frac{\tau_p}{(SF)\gamma S_o} = \frac{2.4}{1.0(62.4)(0.015)} = 2.56 \text{ ft}$$

The allowable depth is less than the depth of the ditch (3 ft).

Step 3. Determine the flow area and hydraulic radius from the geometric properties of a trapezoidal channel for the allowable depth:

$$A = Bd + Zd^2 = 3.3(2.56) + 4(2.56)^2 = 34.7 \text{ ft}^2$$

$$P = B + 2d\sqrt{Z^2 + 1} = 3.3 + 2(2.56)\sqrt{4^2 + 1} = 24.4 \text{ ft}$$

R = A/P = 34.7/24.4 = 1.42 ft

Step 4. From Equation 6.1, n = 0.060

Step 5. Solving Manning's equation (Equation 2.1):

$$Q = \frac{\alpha}{n}(A)(R)^{2/3}(S)^{1/2} = \frac{1.49}{0.060}(34.7)(1.42)^{2/3}(0.015)^{1/2} = 133 \text{ ft}^3/\text{s}$$

Flow in this channel must be limited to the calculated flow by providing a relief inlet or culvert or by providing a lining with greater shear resistance.

This page intentionally left blank.

CHAPTER 4: VEGETATIVE LINING AND BARE SOIL DESIGN

Grass-lined channels have been widely used in roadway drainage systems for many years. They are easily constructed and maintained and work well in a variety of climates and soil conditions. Grass linings provide good erosion protection and can trap sediment and related contaminants in the channel section. Routine maintenance of grass-lined channels consists of mowing, control of weedy plants and woody vegetation, repair of damaged areas and removal of sediment deposits.

The behavior of grass in an open channel lining is complicated by the fact that grass stems bend as flow depth and shear stress increase. This reduces the roughness height and increases velocity and flow rate. For some lining materials (bare earth and rigid linings), the roughness height remains constant regardless of the velocity or depth of flow in the channel. As a result, a grass-lined channel cannot be described by a single roughness coefficient.

The Soil Conservation Service (SCS) (1954) developed a widely used classification of grass channel lining that depends on the degree of retardance. In this classification, retardance is a function of the height and density of the grass cover (USDA, 1987). Grasses are classified into five broad categories, as shown in Table 4.1. Retardance Class A presents the highest resistance to flow and Class E presents the lowest resistance to flow. In general, taller and denser grass species have a higher resistance to flow, while short flexible grasses have a low flow resistance.

Kouwen and Unny (1969) and Kouwen and Li (1981) developed a useful model of the biomechanics of vegetation in open-channel flow. This model provides a general approach for determining the roughness of vegetated channels compared to the retardance classification. The resulting resistance equation (see Appendix C.2) uses the same vegetation properties as the SCS retardance approach, but is more adaptable to the requirements of highway drainage channels. The design approach for grass-lined channels was developed from the Kouwen resistance equation.

Grass linings provide erosion control in two ways. First, the grass stems dissipate shear force within the canopy before it reaches the soil surface. Second, the grass plant (both the root and stem) stabilizes the soil surface against turbulent fluctuations. Temple (SCS, 1954) developed a relationship between the total shear on the lining and the shear at the soil surface based on both processes.

A simple field method is provided to directly measure the density-stiffness parameter of a grass cover. Grass linings for roadside ditches use a wide variety of seed mixes that meet the regional requirements of soil and climate. These seed mix designs are constantly being adapted to improve grass-lined channel performance. Maintenance practices can significantly influence density and uniformity of the grass cover. The sampling of established grasses in roadside ditch application can eliminate much of the uncertainty in lining performance and maintenance practices.

Expertise in vegetation ecology, soil classification, hydrology, and roadway maintenance is required in the design of grass-lined channels. Engineering judgment is essential in determining design parameters based on this expert input. This includes factoring in variations that are unique to a particular roadway design and its operation.

Table 4.1. Retardance Classification of Vegetal Covers

Retardance Class	Cover[1]	Condition
A	Weeping Love Grass	Excellent stand, tall, average 760 mm (30 in)
	Yellow Bluestem Ischaemum	Excellent stand, tall, average 910 mm (36 in)
B	Kudzu	Very dense growth, uncut
	Bermuda Grass	Good stand, tall, average 300 mm (12 in)
	Native Grass Mixture (little bluestem, bluestem, blue gamma, and other long and short midwest grasses)	Good stand, unmowed
	Weeping lovegrass	Good stand, tall, average 610 mm (24 in)
	Lespedeza sericea	Good stand, not woody, tall, average 480 mm (19 in)
	Alfalfa	Good stand, uncut, average 280 mm (11 in)
	Weeping lovegrass	Good stand, unmowed, average 330 mm (13 in)
	Kudzu	Dense growth, uncut
	Blue Gamma	Good stand, uncut, average 280 mm (11 in)
C	Crabgrass	Fair stand, uncut 250 to 1200 mm (10 to 48 in)
	Bermuda grass	Good stand, mowed, average 150 mm (6 in)
	Common Lespedeza	Good stand, uncut, average 280 mm (11 in)
	Grass-Legume mixture--summer (orchard grass, redtop, Italian ryegrass, and common lespedeza)	Good stand, uncut, 150 to 200 mm (6 to 8 in)
	Centipede grass	Very dense cover, average 150 mm (6 in)
	Kentucky Bluegrass	Good stand, headed, 150 to 300 mm (6 to 12 in)
D	Bermuda Grass	Good stand, cut to 60 mm (2.5 in) height
	Common Lespedeza	Excellent stand, uncut, average 110 mm (4.5 in)
	Buffalo Grass	Good stand, uncut, 80 to 150 mm (3 to 6 in)
	Grass-Legume mixture—fall, spring (orchard grass, redtop, Italian ryegrass, and common lespedeza)	Good stand, uncut, 100 to 130 mm (4 to 5 in)
	Lespedeza sericea	After cutting to 50 mm (2 in) height. Very good stand before cutting.
E	Bermuda Grass	Good stand, cut to height, 40 mm (1.5 in)
	Bermuda Grass	Burned stubble

[1] Covers classified have been tested in experimental channels. Covers were green and generally uniform.

4.1 GRASS LINING PROPERTIES

The density, stiffness, and height of grass stems are the main biomechanical properties of grass that relate to flow resistance and erosion control. The stiffness property (product of elasticity and moment of inertia) of grass is similar for a wide range of species (Kouwen, 1988) and is a basic property of grass linings.

Density is the number of grass stems in a given area, i.e., stems per m^2 (ft^2). A good grass lining will have about 2,000 to 4,000 stems/m^2 (200 to 400 stems/ft^2). A poor cover will have about one-third of that density and an excellent cover about five-thirds (USDA, 1987, Table 3.1). While grass density can be determined by physically counting stems, an easier direct method of estimating the density-stiffness property is provided in Appendix E of this manual.

For agricultural ditches, grass heights can reach 0.3 m (1.0 ft) to over 1.0 m (3.3 ft). However, near a roadway grass heights are kept much lower for safety reasons and are typically in the range of 0.075 m (0.25 ft) to 0.225 m (0.75 ft).

The density-stiffness property of grass is defined by the C_s coefficient. C_s can be directly measured using the Fall-Board test (Appendix E) or estimated based on the conditions of the grass cover using Table 4.2. Good cover would be the typical reference condition.

Table 4.2. Density-stiffness Coefficient, C_s

Conditions	Excellent	Very Good	Good	Fair	Poor
C_s (SI)	580	290	106	24	8.6
C_s (CU)	49	25	9.0	2.0	0.73

The combined effect of grass stem height and density-stiffness is defined by the grass roughness coefficient.

$$C_n = \alpha C_s^{0.10} h^{0.528}$$ (4.1)

where,

C_n	=	grass roughness coefficient
C_s	=	density-stiffness coefficient
h	=	stem height, m (ft)
α	=	unit conversion constant, 0.35 (SI), 0.237 (CU)

Table 4.3 provides C_n values for a range of cover and stem height conditions based on Equation 4.1. Denser cover and increased stem height result in increased channel roughness.

Table 4.3. Grass Roughness Coefficient, C_n

Stem Height m (ft)	Excellent	Very Good	Good	Fair	Poor
0.075 (0.25)	0.168	0.157	0.142	0.122	0.111
0.150 (0.50)	0.243	0.227	0.205	0.177	0.159
0.225 (0.75)	0.301	0.281	0.254	0.219	0.197

SCS retardance values relate to a combination of grass stem-height and density. C_n values for standard retardance classes are provided in Table 4.4. Comparing Table 4.3 and 4.4 shows that retardance classes A and B are not commonly found in roadway applications. These retardance classes represent conditions where grass can be allowed to grow much higher than would be permissible for a roadside channel, e.g., wetlands and agricultural ditches. Class E would not be typical of most roadside channel conditions unless they were in a very poor state.

The range of C_n for roadside channels is between 0.10 and 0.30 with a value of 0.20 being common to most conditions and stem heights. In an iterative design process, a good first estimate of the grass roughness coefficient would be $C_n = 0.20$.

Table 4.4 (SI). Grass Roughness Coefficient, C_n, for SCS Retardance Classes

Retardance Class	A	B	C	D	E
Stem Height, mm	910	610	200	100	40
C_s	390	81	47	33	44
C_n	0.605	0.418	0.220	0.147	0.093

Table 4.4 (CU). Grass Roughness Coefficient, C_n, for SCS Retardance Classes

Retardance Class	A	B	C	D	E
Stem Height, in	36	24	8.0	4.0	1.6
C_s	33	7.1	3.9	2.7	3.8
C_n	0.605	0.418	0.220	0.147	0.093

4.2 MANNING'S ROUGHNESS

Manning's roughness coefficient for grass linings varies depending on grass properties as reflected in the C_n parameter and the shear force exerted by the flow. This is because the applied shear on the grass stem causes the stem to bend, which reduces the stem height relative to the depth of flow and reducing the roughness.

$$n = \alpha C_n \tau_o^{-0.4} \tag{4.2}$$

where,

τ_o = mean boundary shear stress, N/m^2 (lb/ft^2)

α = unit conversion constant, 1.0 (SI), 0.213 (CU)

See Appendix C.2 for the derivation of Equation 4.2.

4.3 PERMISSIBLE SHEAR STRESS

The permissible shear stress of a vegetative lining is determined both by the underlying soil properties as well as those of the vegetation. Determination of permissible shear stress for the lining is based on the permissible shear stress of the soil combined with the protection afforded by the vegetation, if any.

4.3.1 Effective Shear Stress

Grass lining moves shear stress away from the soil surface. The remaining shear at the soil surface is termed the effective shear stress. When the effective shear stress is less than the allowable shear for the soil surface, then erosion of the soil surface will be controlled. Grass linings provide shear reduction in two ways. First, the grass stems dissipate shear force within the canopy before it reaches the soil surface. Second, the grass plant (both the root and stem) stabilizes the soil surface against turbulent fluctuations. This process model (USDA, 1987) for the effective shear at the soil surface is given by the following equation.

$$\tau_e = \tau_d (1 - C_f) \left(\frac{n_s}{n} \right)^2 \tag{4.3}$$

where,

τ_e = effective shear stress on the soil surface, N/m^2 (lb/ft^2)

τ_d = design shear stress, N/m^2 (lb/ft^2)

C_f = grass cover factor

n_s = soil grain roughness

n = overall lining roughness

Soil grain roughness is taken as 0.016 when $D_{75} < 1.3$ mm (0.05 in). For larger grain soils, the soil grain roughness is given by:

$$n_s = \alpha (D_{75})^{\frac{1}{6}}$$

(4.4)

where,

n_s = soil grain roughness ($D_{75} > 1.3$ mm (0.05 in))

D_{75} = soil size where 75% of the material is finer, mm (in)

α = unit conversion constant, 0.015 (SI), 0.026 (CU)

Note that soil grain roughness value, n_s, is less than the typical value reported in Table 2.1 for a bare soil channel. The total roughness value for bare soil channel includes form roughness (surface texture of the soil) in addition to the soil grain roughness. However, Equation 4.3 is based on soil grain roughness.

The grass cover factor, C_f, varies with cover density and grass growth form (sod or bunch). The selection of the cover factor is a matter of engineering judgment since limited data are available. Table 4.5 provides a reasonable approach to estimating a cover factor based on (USDA, 1987, Table 3.1). Cover factors are better for sod-forming grasses than bunch grasses. In all cases a uniform stand of grass is assumed. Non-uniform conditions include wheel ruts, animal trails and other disturbances that run parallel to the direction of the channel. Estimates of cover factor are best for good uniform stands of grass and there is more uncertainty in the estimates of fair and poor conditions.

Table 4.5. Cover Factor Values for Uniform Stands of Grass

Growth Form	Cover Factor, C_f				
	Excellent	Very Good	Good	Fair	Poor
Sod	0.98	0.95	0.90	0.84	0.75
Bunch	0.55	0.53	0.50	0.47	0.41
Mixed	0.82	0.79	0.75	0.70	0.62

4.3.2 Permissible Soil Shear Stress

Erosion of the soil boundary occurs when the effective shear stress exceeds the permissible soil shear stress. Permissible soil shear stress is a function of particle size, cohesive strength, and soil density. The erodibility of coarse non-cohesive soils (defined as soils with a plasticity index of less than 10) is due mainly to particle size, while fine-grained cohesive soils are controlled mainly by cohesive strength and soil density.

New ditch construction includes the placement of topsoil on the perimeter of the channel. Topsoil is typically gathered from locations on the project and stockpiled for revegetation work. Therefore, the important physical properties of the soil can be determined during the design by sampling surface soils from the project area. Since these soils are likely to be mixed together, average physical properties are acceptable for design.

The following sections offer detailed methods for determination of soil permissible shear. However, the normal variation of permissible shear stress for different soils is moderate, particularly for fine-grained cohesive soils. An approximate method is also provided for cohesive soils.

4.3.2.1 Non-cohesive Soils

The permissible soil shear stress for fine-grained, non-cohesive soils (D_{75} < 1.3 mm (0.05 in)) is relatively constant and is conservatively estimated at 1.0 N/m^2 (0.02 lb/ft^2). For coarse grained, non-cohesive soils (1.3 mm (0.05 in) < D_{75} < 50 mm (2 in)) the following equation applies.

$$\tau_{p,soil} = \alpha D_{75}$$ (4.5)

where,

$\tau_{p,soil}$ = permissible soil shear stress, N/m^2 (lb/ft^2)

D_{75} = soil size where 75% of the material is finer, mm (in)

α = unit conversion constant, 0.75 (SI), 0.4 (CU)

4.3.2.2 Cohesive Soils

Cohesive soils are largely fine grained and their permissible shear stress depends on cohesive strength and soil density. Cohesive strength is associated with the plasticity index (PI), which is the difference between the liquid and plastic limits of the soil. The soil density is a function of the void ratio (e). The basic formula for permissible shear on cohesive soils is the following.

$$\tau_{p,soil} = \left(c_1 PI^2 + c_2 PI + c_3\right)\left(c_4 + c_5 e\right)^2 c_6$$ (4.6)

where,

$\tau_{p,soil}$ = soil permissible shear stress, N/m^2 (lb/ft^2)

PI = plasticity index

e = void ratio

c_1, c_2, c_3, c_4, c_5, c_6 = coefficients (Table 4.6)

A simplified approach for estimating permissible soil shear stress based on Equation 4.6 is illustrated in Figure 4.1. Fine grained soils are grouped together (GM, CL, SC, ML, SM, and MH) and coarse grained soil (GC). Clays (CH) fall between the two groups.

Higher soil unit weight increases the permissible shear stress and lower soil unit weight decreases permissible shear stress. Figure 4.1 is applicable for soils that are within 5 percent of a typical unit weight for a soil class. For sands and gravels (SM, SC, GM, GC) typical soil unit weight is approximately 1.6 ton/m^3 (100 lb/ft^3), for silts and lean clays (ML, CL) 1.4 ton/m^3 (90 lb/ft^3) and fat clays (CH, MH) 1.3 ton/m^3 (80 lb/ft^3).

Table 4.6. Coefficients for Permissible Soil Shear Stress (USDA, 1987)

ASTM Soil Classification[1]	Applicable Range	c_1	c_2	c_3	c_4	c_5	c_6 (SI)	c_6 (CU)
GM	$10 \le PI \le 20$	1.07	14.3	47.7	1.42	-0.61	4.8×10^{-3}	10^{-4}
	$20 \le PI$			0.076	1.42	-0.61	48.	1.0
GC	$10 \le PI \le 20$	0.0477	2.86	42.9	1.42	-0.61	4.8×10^{-2}	10^{-3}
	$20 \le PI$			0.119	1.42	-0.61	48.	1.0
SM	$10 \le PI \le 20$	1.07	7.15	11.9	1.42	-0.61	4.8×10^{-3}	10^{-4}
	$20 \le PI$			0.058	1.42	-0.61	48.	1.0
SC	$10 \le PI \le 20$	1.07	14.3	47.7	1.42	-0.61	4.8×10^{-3}	10^{-4}
	$20 \le PI$			0.076	1.42	-0.61	48.	1.0
ML	$10 \le PI \le 20$	1.07	7.15	11.9	1.48	-0.57	4.8×10^{-3}	10^{-4}
	$20 \le PI$			0.058	1.48	-0.57	48.	1.0
CL	$10 \le PI \le 20$	1.07	14.3	47.7	1.48	-0.57	4.8×10^{-3}	10^{-4}
	$20 \le PI$			0.076	1.48	-0.57	48.	1.0
MH	$10 \le PI \le 20$	0.0477	1.43	10.7	1.38	-0.373	4.8×10^{-2}	10^{-3}
	$20 \le PI$			0.058	1.38	-0.373	48.	1.0
CH	$20 \le PI$			0.097	1.38	-0.373	48.	1.0

(1) Note: Typical names

GM   Silty gravels, gravel-sand silt mixtures

GC   Clayey gravels, gravel-sand-clay mixtures

SM   Silty sands, sand-silt mixtures

SC   Clayey sands, sand-clay mixtures

ML   Inorganic silts, very fine sands, rock flour, silty or clayey fine sands

CL   Inorganic clays of low to medium plasticity, gravelly clays, sandy clays, silty clays, lean clays

MH   Inorganic silts, micaceous or diatomaceous fine sands or silts, elastic silts

CH   Inorganic clays of high plasticity, fat clays

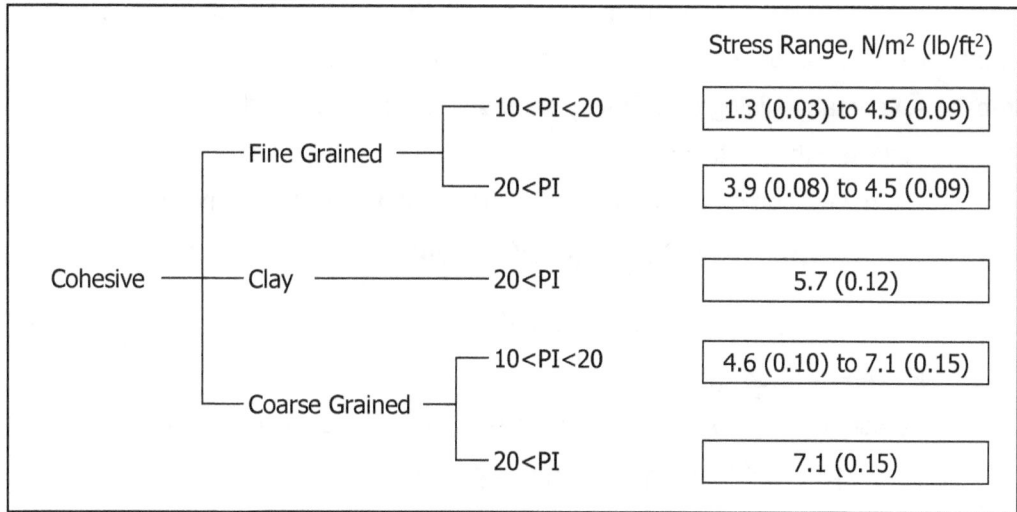

Figure 4.1. Cohesive Soil Permissible Shear Stress

4.3.3 Permissible Vegetation/Soil Shear Stress

The combined effects of the soil permissible shear stress and the effective shear stress transferred through the vegetative lining results in a permissible shear stress for the vegetative lining. Taking Equation 4.3 and substituting the permissible shear stress for the soil for the effective shear stress on the soil, τ_e, gives the following equation for permissible shear stress for the vegetative lining:

$$\tau_p = \frac{\tau_{p,soil}}{(1-C_f)}\left(\frac{n}{n_s}\right)^2 \qquad (4.7)$$

where,

τ_p	=	permissible shear stress on the vegetative lining, N/m^2 (lb/ft^2)
$\tau_{p,soil}$	=	permissible soil shear stress, N/m^2 (lb/ft^2)
C_f	=	grass cover factor
n_s	=	soil grain roughness
n	=	overall lining roughness

Design Example: Grass Lining Design (SI)

Evaluate a grass lining for a roadside channel given the following channel shape, soil conditions, grade, and design flow. It is expected that the grass lining will be maintained in good conditions in the spring and summer months, which are the main storm seasons.

Given:

Shape: Trapezoidal, B = 0.9 m, Z = 3
Soil: Clayey sand (SC classification), PI = 16, e = 0.5
Grass: Sod, height = 0.075 m
Grade: 3.0 percent
Flow: 0.5 m^3/s

Solution

The solution is accomplished using procedure given in Section 3.1 for a straight channel.

Step 1. Channel slope, shape, and discharge have been given.

Step 2. A vegetative lining on a clayey sand soil will be evaluated.

Step 3. Initial depth is estimated at 0.30 m

From the geometric relationship of a trapezoid (see Appendix B):

$$A = Bd + Zd^2 = 0.9(0.3) + 3(0.3)^2 = 0.540 \text{ m}^2$$

$$P = B + 2d\sqrt{Z^2 + 1} = 0.9 + 2(0.3)\sqrt{3^2 + 1} = 2.80 \text{ m}$$

R = A/P = (0.54)/(2.8) = 0.193 m

Step 4. To estimate n, the applied shear stress on the grass lining is given by Equation 2.3

$$\tau_o = \gamma \, RS_o = 9810(0.193)(0.03) = 56.8 \text{ N/m}^2$$

Determine a Manning's n value from Equation 4.2. From Table 4.3, C_n = 0.142

$$n = \alpha C_n \tau^{-0.4} = 1.0(0.142)(56.8)^{-0.4} = 0.028$$

The discharge is calculated using Manning's equation (Equation 2.1):

$$Q = \frac{\alpha}{n} AR^{\frac{2}{3}} S_f^{\frac{1}{2}} = \frac{1}{0.028}(0.540)(0.193)^{\frac{2}{3}}(0.03)^{\frac{1}{2}} = 1.12 \, m^3/s$$

Step 5. Since this value is more than 5 percent different from the design flow, we need to go back to step 3 to estimate a new flow depth.

Step 3 (2nd iteration). Estimate a new depth solving Equation 2.2 or other appropriate method iteratively to find the next estimate for depth:

d = 0.21 m

Revise the hydraulic radius.

$$A = Bd + Zd^2 = 0.9(0.21) + 3(0.21)^2 = 0.321 \, m^2$$

$$P = B + 2d\sqrt{Z^2 + 1} = 0.9 + 2(0.21)\sqrt{3^2 + 1} = 2.23 \, m$$

R = A/P = (0.321)/(2.23) = 0.144 m

Step 4 (2nd iteration). To estimate n, the applied shear stress on the grass lining is given by Equation 2.3

$$\tau_o = \gamma R S_o = 9810(0.144)(0.03) = 42.4 \, N/m^2$$

Determine a Manning's n value from Equation 4.2. From Table 4.3, C_n = 0.142

$$n = \alpha C_n \tau^{-0.4} = 1.0(0.142)(42.4)^{-0.4} = 0.032$$

The discharge is calculated using Manning's equation (Equation 2.1):

$$Q = \frac{\alpha}{n} AR^{\frac{2}{3}} S_f^{\frac{1}{2}} = \frac{1}{0.032}(0.321)(0.144)^{\frac{2}{3}}(0.03)^{\frac{1}{2}} = 0.48 \, m^3/s$$

Step 5 (2nd iteration). Since this value is within 5 percent of the design flow, we can proceed to step 6.

Step 6. The maximum shear on the channel bottom is.

$$\tau_d = \gamma d S_o = 9810(0.21)(0.03) = 61.8 \, N/m^2$$

Determine the permissible soil shear stress from Equation 4.6.

$$\tau_{p,soil} = (c_1 PI^2 + c_2 PI + c_3)(c_4 + c_5 e)^2 c_6 = (1.07(16)^2 + 14.3(16) + 47.7)(1.42 - 0.61(0.5))^2(0.0048) = 3.28 \, N/m^2$$

Equation 4.7 gives the permissible shear stress on the vegetation. The value of C_f is found in Table 4.5.

$$\tau_p = \frac{\tau_{p,soil}}{(1 - C_f)}\left(\frac{n}{n_s}\right)^2 = \frac{3.28}{(1 - 0.9)}\left(\frac{0.032}{0.016}\right)^2 = 131 \, N/m^2$$

The safety factor for this channel is taken as 1.0.

Step 7. The grass lining is acceptable since the maximum shear on the vegetation is less than the permissible shear of 131 N/m^2.

Design Example: Grass Lining Design (CU)

Evaluate a grass lining for a roadside channel given the following channel shape, soil conditions, grade, and design flow. It is expected that the grass lining will be maintained in good conditions in the spring and summer months, which are the main storm seasons.

Given:
Shape: Trapezoidal, B = 3.0 ft, Z = 3
Soil: Clayey sand (SC classification), PI = 16, e = 0.5
Grass: Sod, height = 0.25 ft
Grade: 3.0 percent
Flow: 17.5 ft^3/s

Solution

The solution is accomplished using procedure given in Section 3.1 for a straight channel.

Step 1. Channel slope, shape, and discharge have been given.

Step 2. A vegetative lining on a clayey sand soil will be evaluated.

Step 3. Initial depth is estimated at 1.0 ft.

From the geometric relationship of a trapezoid (see Appendix B):

$$A = Bd + Zd^2 = 3.0(1.0) + 3(1.0)^2 = 6.00 \text{ ft}^2$$

$$P = B + 2d\sqrt{Z^2 + 1} = 3.0 + 2(1.0)\sqrt{3^2 + 1} = 9.32 \text{ ft}$$

R = A/P = (6.00)/(9.32) = 0.643 ft

Step 4. To estimate n, the applied shear stress on the grass lining is given by Equation 2.3

$$\tau_o = \gamma RS_o = 62.4(0.643)(0.03) = 1.20 \text{ lb / ft}^2$$

Determine a Manning's n value from Equation 4.2. From Table 4.3, C_n = 0.142

$$n = \alpha C_n \tau^{-0.4} = 0.213(0.142)(1.20)^{-0.4} = 0.028$$

The discharge is calculated using Manning's equation (Equation 2.1):

$$Q = \frac{\alpha}{n} AR^{2/3}S_f^{1/2} = \frac{1.49}{0.028}(6.00)(0.643)^{2/3}(0.03)^{1/2} = 41.2 \text{ ft}^3 / s$$

Step 5. Since this value is more than 5 percent different from the design flow, we need to go back to step 3 to estimate a new flow depth.

Step 3 (2nd iteration). Estimate a new depth solving Equation 2.2 or other appropriate method iteratively to find the next estimate for depth:

d = 0.70 ft

Revise the hydraulic radius.

$$A = Bd + Zd^2 = 3.0(0.70) + 3(0.70)^2 = 3.57 \text{ ft}^2$$

$$P = B + 2d\sqrt{Z^2 + 1} = 3.0 + 2(0.70)\sqrt{3^2 + 1} = 7.43 \text{ ft}$$

$$R = A/P = (3.57)/(7.43) = 0.481 \text{ ft}$$

Step 4 (2nd iteration). To estimate n, the applied shear stress on the grass lining is given by Equation 2.3

$$\tau_o = \gamma R S_o = 62.4(0.481)(0.03) = 0.90 \text{ lb} / \text{ft}^2$$

Determine a Manning's n value from Equation 4.2. From Table 4.3, $C_n = 0.142$

$$n = \alpha C_n \tau^{-0.4} = 0.213(0.142)(0.90)^{-0.4} = 0.032$$

The discharge is calculated using Manning's equation (Equation 2.1):

$$Q = \frac{\alpha}{n} AR^{2/3}S_f^{1/2} = \frac{1.49}{0.032}(3.57)(0.481)^{2/3}(0.03)^{1/2} = 17.7 \text{ ft}^3 / \text{s}$$

Step 5 (2nd iteration). Since this value is within 5 percent of the design flow, we can proceed to step 6.

Step 6. The maximum shear on the channel bottom is.

$$\tau_d = \gamma dS_o = 62.4(0.70)(0.03) = 1.31 \text{ lb} / \text{ft}^2$$

Determine the permissible soil shear stress from Equation 4.6.

$$\tau_{p,soil} = (c_1 PI^2 + c_2 PI + c_3)(c_4 + c_5 e)^2 c_6 = (1.07(16)^2 + 14.3(16) + 47.7)(1.42 - 0.61(0.5))^2(0.0001) = 0.068 \text{ lb} / \text{ft}^2$$

Equation 4.7 gives the permissible shear stress on the vegetation. The value of C_f is found in Table 4.5.

$$\tau_p = \frac{\tau_{p,soil}}{(1 - C_f)}\left(\frac{n}{n_s}\right)^2 = \frac{0.068}{(1 - 0.9)}\left(\frac{0.032}{0.016}\right)^2 = 2.7 \text{ lb} / \text{ft}^2$$

The safety factor for this channel is taken as 1.0.

Step 7. The grass lining is acceptable since the maximum shear on the vegetation is less than the permissible shear of 2.7 lb/ft^2.

4.4 MAXIMUM DISCHARGE APPROACH

The maximum discharge for a vegetative lining is estimated following the basic steps outlined in Section 3.6. To accomplish this, it is necessary to develop a means of estimating the applied bottom shear stress that will yield the permissible effective shear stress on the soil. Substituting Equation 4.2 into Equation 4.3 and assuming the $\tau_o = 0.75 \tau_d$ and solving for τ_d yields:

$$\tau_d = \left[\frac{\alpha \tau_e}{(1 - C_f)}\left(\frac{C_n}{n_s}\right)^2\right]^{5/9} \qquad (4.8)$$

where,

α = unit conversion constant, 1.26 (SI), 0.057 (CU)

The assumed relationship between τ_o and τ_d is not constant. Therefore, once the depth associated with maximum discharge has been found, a check should be conducted to verify the assumption.

Design Example: Maximum Discharge for a Grass Lining (SI)

Determine the maximum discharge for a grass-lined channel given the following shape, soil conditions, and grade.

Given:

Shape: Trapezoidal, B = 0.9 m, z = 3

Soil: Silty sand (SC classification), PI = 5, D_{75} = 2 mm

Grade: 5.0 percent

Solution

The solution is accomplished using procedure given in Section 3.6 for a maximum discharge approach.

Step 1. The candidate lining is a sod forming grass in good condition with a stem height of 0.150 m.

Step 2. Determine the maximum depth. For a grass lining this requires several steps. First, determine the permissible soil shear stress. From Equation 4.5:

$$\tau_p = \alpha D_{75} = 0.75(2) = 1.5 \, N/m^2$$

To estimate the shear, we will first need to use Equation 4.1 to estimate C_n with C_s taken from Table 4.2

$$C_n = \alpha C_s^{0.10} h^{0.528} = 0.35(106)^{0.1}(0.150)^{0.528} = 0.205$$

Next, estimate the maximum applied shear using Equation 4.8.

$$\tau_d = \left[\frac{\alpha \tau_e}{(1-C_f)}\left(\frac{C_n}{n_s}\right)^2\right]^{\frac{5}{9}} = \left[\frac{1.25(1.5)}{(1-0.9)}\left(\frac{0.205}{0.016}\right)^2\right]^{\frac{5}{9}} = 87 \, N/m^2$$

Maximum depth from Equation 3.10 with a safety factor of 1.0 is:

$$d = \frac{\tau_d}{(SF)\gamma S_o} = \frac{87}{(1.0)9800(0.05)} = 0.18 \, m$$

Step 3. Determine the area and hydraulic radius corresponding to the allowable depth based on the channel geometry

$$A = Bd + Zd^2 = 0.90(0.18) + 3(0.18)^2 = 0.259 \, m^2$$

$$P = B + 2d\sqrt{Z^2 + 1} = 0.9 + 2(0.18)\sqrt{3^2 + 1} = 2.04 \, m$$

R = A/P = (0.259)/(2.04) = 0.127 m

Step 4. Estimate the Manning's n value appropriate for the lining type from Equation 4.2, but first calculate the mean boundary shear.

$$\tau_o = \gamma\, RS_o = 9810(0.127)(0.05) = 62.3\ N/m^2$$

$$n = \alpha C_n \tau_o^{-0.4} = 1.0(0.205)(62.3)^{-0.4} = 0.039$$

Step 5. Solve Manning's equation to determine the maximum discharge for the channel.

$$Q = \frac{\alpha}{n} AR^{\frac{2}{3}} S^{\frac{1}{2}} = \frac{1}{0.039}(0.259)(0.127)^{\frac{2}{3}}(0.05)^{\frac{1}{2}} = 0.38\ m^3/s$$

Since Equation 4.8 used in Step 2 is an approximate equation, check the effective shear stress using Equation 4.3.

$$\tau_e = \tau_d\!\left(1 - C_f\right)\!\left(\frac{n_s}{n}\right)^2 = 87(1 - 0.9)\!\left(\frac{0.016}{0.039}\right)^2 = 1.46\ N/m^2$$

Since this value is less than, but close to τ_p for the soil 1.5 N/m^2, the maximum discharge is 0.38 m^3/s.

Design Example: Maximum Discharge for a Grass Lining (CU)

Determine the maximum discharge for a grass-lined channel given the following shape, soil conditions, and grade.

Given:
  Shape: Trapezoidal, B = 3.0 ft, z = 3
  Soil: Silty sand (SC classification), PI = 5, D_{75} = 0.08 in
  Grade: 5.0 percent

Solution

The solution is accomplished using procedure given in Section 3.6 for a maximum discharge approach.

Step 1. The candidate lining is a sod forming grass in good condition with a stem height of 0.5 ft.

Step 2. Determine the maximum depth. For a grass lining this requires several steps. First, determine the permissible soil shear stress. From Equation 4.5:

$$\tau_p = \alpha D_{75} = 0.4(0.08) = 0.032\ lb/ft^2$$

To estimate the shear, we will first need to use Equation 4.1 to estimate C_n with C_s taken from Table 4.2

$$C_n = \alpha C_s^{0.10} h^{0.528} = 0.237(9.0)^{0.1}(0.5)^{0.528} = 0.205$$

Next, estimate the maximum applied shear using Equation 4.8.

$$\tau_d = \left[\frac{\alpha \tau_e}{\left(1 - C_f\right)}\left(\frac{C_n}{n_s}\right)^2\right]^{\frac{5}{9}} = \left[\frac{0.057(0.032)}{(1 - 0.9)}\left(\frac{0.205}{0.016}\right)^2\right]^{\frac{5}{9}} = 1.84\ lb/ft^2$$

4-13

Maximum depth from Equation 3.10 with a safety factor of 1.0 is:

$$d = \frac{\tau_d}{(SF)\gamma S_o} = \frac{1.84}{(1.0)62.4(0.05)} = 0.59 \text{ ft}$$

Step 3. Determine the area and hydraulic radius corresponding to the allowable depth based on the channel geometry

$$A = Bd + Zd^2 = 3.0(0.59) + 3(0.59)^2 = 2.81 \text{ ft}^2$$

$$P = B + 2d\sqrt{Z^2 + 1} = 3.0 + 2(0.59)\sqrt{3^2 + 1} = 6.73 \text{ ft}$$

$$R = A/P = (2.81)/(6.73) = 0.42 \text{ ft}$$

Step 4. Estimate the Manning's n value appropriate for the lining type from Equation 4.2, but first calculate the mean boundary shear.

$$\tau_o = \gamma R S_o = 62.4(0.42)(0.05) = 1.31 \text{ lb / ft}^2$$

$$n = \alpha C_n \tau_o^{-0.4} = 0.213(0.205)(1.31)^{-0.4} = 0.039$$

Step 5. Solve Manning's equation to determine the maximum discharge for the channel.

$$Q = \frac{\alpha}{n} A R^{2/3} S^{1/2} = \frac{1.49}{0.039}(2.81)(0.42)^{2/3}(0.05)^{1/2} = 13.5 \text{ ft}^3 / s$$

Since Equation 4.8 used in Step 2 is an approximate equation, check the effective shear stress using Equation 4.3.

$$\tau_e = \tau_d\left(1 - C_f\right)\left(\frac{n_s}{n}\right)^2 = 1.84(1 - 0.9)\left(\frac{0.016}{0.039}\right)^2 = 0.031 \text{ lb / ft}^2$$

Since this value is less than, but close to τ_p for the soil 0.032 lb/ft^2, the maximum discharge is 13.5 ft^3/s.

4.5 TURF REINFORCEMENT WITH GRAVEL/SOIL MIXTURE

The rock products industry provides a variety of uniformly graded gravels for use as mulch and soil stabilization. A gravel/soil mixture provides a non-degradable lining that is created as part of the soil preparation and is followed by seeding. The integration of gravel and soil is accomplished by mixing (by raking or disking the gravel into the soil). The gravel provides a matrix of sufficient thickness and void space to permit establishment of vegetation roots within the matrix. It provides enhanced erosion resistance during the vegetative establishment period and it provides a more resistant underlying layer than soil once vegetation is established.

The density, size and gradation of the gravel are the main properties that relate to flow resistance and erosion control performance. Stone specific gravity should be approximately 2.6 (typical of most stone). The stone should be hard and durable to ensure transport without breakage. Placed density of uniformly graded gravel is 1.76 metric ton/m^3 (1.5 ton/yd^3). A uniform gradation is necessary to permit germination and growth of grass plants through the gravel layer. Table 4.7 provides two typical gravel gradations for use in erosion control.

Table 4.7. Gravel Gradation Table, Percentages Passing Nominal Size Designations

Size	Very Coarse (D_{75} = 45 mm (1.75 in))	Coarse (D_{75} = 30 mm (1.2 in))
50.0 mm (2 in)	90 - 100	
37.5 mm (1.5 in)	35 - 70	90 – 100
25.0 mm (1 in)	0 - 15	35 – 70
19.0 mm (0.75 in)		0 – 15

The application rate of gravel mixed into the soil should result in 25 percent of the mixture in the gravel size. Generally, soil preparation for a channel lining will be to a depth of 75 to 100 mm (3 to 4 inches). The application rate of gravel to the prepared soil layer that results in a 25 percent gravel mix is calculated as follows.

$$I_{gravel} = \alpha \left(\frac{1 - i_{gravel}}{3} \right) T_s \, \gamma_{gravel} \qquad (4.9)$$

where,

I_{gravel} = gravel application rate, metric ton/m^2 (ton/yd^2)

i_{gravel} = fraction of gravel (equal to or larger than gravel layer size) already in the soil

T_s = thickness of the soil surface, m (ft)

γ_{gravel} = unit weight of gravel, metric ton/m^3 (ton/yd^3)

α = unit conversion constant, 1.0 (SI), 0.333 (CU)

The gravel application rates for fine-grained soils (i_{gravel} = 0) are summarized in Table 4.8. If the soil already contains some coarse gravel, then the application rate can be reduced by 1- i_{gravel}.

Table 4.8. Gravel Application Rates for Fine Grain Soils

Soil Preparation Depth	Application Rate, I_{gravel}
75 mm (3 inches)	0.044 ton/m^2 (0.041 ton/yd^2)
100 mm (4 inches)	0.058 ton/m^2 (0.056 ton/yd^2)

The effect of roadside maintenance activities, particularly mowing, on longevity of gravel/soil mixtures needs to be considered. Gravel/soil linings are unlikely to be displaced by mowing since they are heavy. They are also a particle-type lining, so loss of a few stones will not affect overall lining integrity. Therefore, a gravel/soil mix is a good turf reinforcement alternative.

Design Example: Turf Reinforcement with a Gravel/Soil Mixture (SI)

Evaluate the following proposed lining design for a vegetated channel reinforced with a coarse gravel soil amendment. The gravel will be mixed into the soil to result in 25 percent gravel. Since there is no existing gravel in the soil, an application rate of 0.058 ton/m^2 is recommended (100 mm soil preparation depth). See Table 4.8.

Given:

     Shape: Trapezoidal, B = 0.9 m, Z = 3

     Soil: Silty sand (SC classification), PI = 5, D_{75} = 2 mm

     Grass: Sod, good condition, h = 0.150 m

     Gravel: D_{75} = 25 mm

     Grade: 5.0 percent

     Flow: 1.7 m^3/s

Solution

The solution is accomplished using procedure given in Section 3.1 for a straight channel.

    Step 1. Channel slope, shape, and discharge have been given.

    Step 2. Proposed lining is a vegetated channel with a gravel soil amendment.

    Step 3. Initial depth is estimated at 0.30 m

            From the geometric relationship of a trapezoid (see Appendix B):

$$A = Bd + Zd^2 = 0.9(0.3) + 3(0.3)^2 = 0.540 \text{ m}^2$$

$$P = B + 2d\sqrt{Z^2 + 1} = 0.9 + 2(0.3)\sqrt{3^2 + 1} = 2.80 \text{ m}$$

R = A/P = (0.540 m^2)/(2.80 m) = 0.193 m

    Step 4. To estimate n, the applied shear stress on the grass lining is given by Equation 2.3

$$\tau_o = \gamma RS_o = 9810(0.193)(0.05) = 94.7 \text{ N/m}^2$$

            Determine a Manning's n value from Equation 4.2. From Table 4.3, C_n = 0.205

$$n = \alpha C_n \tau^{-0.4} = 1.0(0.205)(94.7)^{-0.4} = 0.033$$

            The discharge is calculated using Manning's equation (Equation 2.1):

$$Q = \frac{\alpha}{n} AR^{2/3}S_f^{1/2} = \frac{1}{0.033}(0.540)(0.193)^{2/3}(0.05)^{1/2} = 1.22 \text{ m}^3/s$$

    Step 5. Since this value is more than 5 percent different from the design flow, we need to go back to step 3 to estimate a new flow depth.

    Step 3 (2nd iteration). Estimate a new depth solving Equation 2.2 or other appropriate method iteratively to find the next estimate for depth:

            d = 0.35 m

            Revise hydraulic radius.

$$A = Bd + Zd^2 = 0.9(0.35) + 3(0.35)^2 = 0.682 \text{ m}^2$$

$$P = B + 2d\sqrt{Z^2 + 1} = 0.9 + 2(0.35)\sqrt{3^2 + 1} = 3.11 \text{ m}$$

R = A/P = (0.682)/(3.11) = 0.219 m

    Step 4 (2nd iteration). To estimate n, the applied shear stress on the grass lining is given by Equation 2.3

$$\tau_o = \gamma \, RS_o = 9810(0.219)(0.05) = 107 \, N/m^2$$

Determine a Manning's n value for the vegetation from Equation 4.2. From Table 4.3, $C_n = 0.142$

$$n = \alpha C_n \tau^{-0.4} = 1(0.205)(107)^{-0.4} = 0.032$$

The discharge is calculated using Manning's equation (Equation 2.1):

$$Q = \frac{\alpha}{n} AR^{2/3} S_f^{1/2} = \frac{1}{0.032}(0.682)(0.219)^{2/3}(0.05)^{1/2} = 1.73 \, m^3/s$$

Step 5 (2nd iteration). Since this value is within 5 percent of the design flow, we can proceed to step 6.

Step 6. The maximum shear on the channel bottom is.

$$\tau_d = \gamma dS_o = 9810(0.35)(0.05) = 172 \, N/m^2$$

Determine the permissible shear stress from Equation 4.4. For turf reinforcement with gravel/soil the D_{75} for the gravel is used instead of the D_{75} for the soil.

$$\tau_{p,soil} = \alpha D_{75} = 0.75(25) = 19 \, N/m^2$$

A Manning's n for the soil/gravel mixture is derived from Equation 4.4:

$$n_s = \alpha D_{75}^{1/6} = 0.015(25)^{1/6} = 0.026$$

Equation 4.7 gives the permissible shear stress on the vegetation. The value of C_f is found in Table 4.5.

$$\tau_p = \frac{\tau_{p,soil}}{(1-C_f)}\left(\frac{n}{n_s}\right)^2 = \frac{19}{(1-0.9)}\left(\frac{0.032}{0.026}\right)^2 = 288 \, N/m^2$$

The safety factor for this channel is taken as 1.0.

Step 7. The grass lining reinforced with the gravel/soil mixture is acceptable since the permissible shear is greater than the maximum shear

Design Example: Turf Reinforcement with a Gravel/Soil Mixture (CU)

Evaluate the following proposed lining design for a vegetated channel reinforced with a coarse gravel soil amendment. The gravel will be mixed into the soil to result in 25 percent gravel. Since there is no gravel in the soil, an application rate of 0.056 ton/yd^2 is recommended (4 inch soil preparation depth). See Table 4.8.

Given:

Shape: Trapezoidal, B = 3 ft, Z = 3

Soil: Silty sand (SC classification), PI = 5, D_{75} = 0.08 in

Grass: sod, good condition, h = 0.5 in

Gravel: D_{75} = 1.0 in

Grade: 5.0 percent

Flow: 60 ft^3/s

Solution

The solution is accomplished using procedure given in Section 3.1 for a straight channel.

   Step 1. Channel slope, shape, and discharge have been given.

   Step 2. Proposed lining is a vegetated channel with a gravel soil amendment.

   Step 3. Initial depth is estimated at 1.0 ft

From the geometric relationship of a trapezoid (see Appendix B):

$$A = Bd + Zd^2 = 3.0(1) + 3(1.0)^2 = 6.00 \text{ ft}^2$$

$$P = B + 2d\sqrt{Z^2 + 1} = 3.0 + 2(1.0)\sqrt{3^2 + 1} = 9.32 \text{ ft}$$

$$R = A/P = (6.00)/(9.32) = 0.644 \text{ ft}$$

   Step 4. To estimate n, the applied shear stress on the grass lining is given by Equation 2.3

$$\tau_o = \gamma\, RS_o = 62.4(0.644)(0.05) = 2.01 \text{ lb} / \text{ft}^2$$

Determine a Manning's n value from Equation 4.2. From Table 4.3, $C_n = 0.205$

$$n = \alpha C_n \tau^{-0.4} = 0.213(0.205)(2.01)^{-0.4} = 0.033$$

The discharge is calculated using Manning's equation (Equation 2.1):

$$Q = \frac{\alpha}{n} AR^{2/3}S_f^{1/2} = \frac{1.49}{0.033}(6.00)(0.644)^{2/3}(0.05)^{1/2} = 45.2 \text{ ft}^3 / \text{s}$$

   Step 5. Since this value is more than 5 percent different from the design flow, we need to go back to step 3 to estimate a new flow depth.

   Step 3 (2nd iteration). Estimate a new depth solving Equation 2.2 or other appropriate method iteratively to find the next estimate for depth:

d = 1.13 ft

Revise hydraulic radius.

$$A = Bd + Zd^2 = 3.0(1.13) + 3(1.13)^2 = 7.22 \text{ ft}^2$$

$$P = B + 2d\sqrt{Z^2 + 1} = 3.0 + 2(1.13)\sqrt{3^2 + 1} = 10.1 \text{ ft}$$

$$R = A/P = (7.22)/(10.1) = 0.715 \text{ ft}$$

   Step 4 (2nd iteration). To estimate n, the applied shear stress on the grass lining is given by Equation 2.3

$$\tau_o = \gamma\, RS_o = 62.4(0.715)(0.05) = 2.23 \text{ lb} / \text{ft}^2$$

Determine a Manning's n value from Equation 4.2. From Table 4.3, $C_n = 0.205$

$$n = \alpha C_n \tau^{-0.4} = 0.213(0.205)(2.23)^{-0.4} = 0.032$$

The discharge is calculated using Manning's equation (Equation 2.1):

$$Q = \frac{\alpha}{n} AR^{2/3}S_f^{1/2} = \frac{1.49}{0.032}(7.22)(0.715)^{2/3}(0.05)^{1/2} = 60.1 \text{ ft}^3 / \text{s}$$

Step 5 (2nd iteration). Since this value is within 5 percent of the design flow, we can proceed to step 6.

Step 6. The maximum shear on the channel bottom is.

$$\tau_d = \gamma d S_o = 62.4(1.13)(0.05) = 3.53 \text{ lb / ft}^2$$

Determine the permissible shear stress from Equation 4.4. For turf reinforcement with gravel/soil the D_{75} for the gravel is used instead of the D_{75} for the soil.

$$\tau_{p,soil} = \alpha D_{75} = 0.4(1.0) = 0.4 \text{ lb / ft}^2$$

A Manning's n for the soil/gravel mixture is derived from Equation 4.4:

$$n_s = \alpha D_{75}^{1/6} = 0.026(1.0)^{1/6} = 0.026$$

Equation 4.7 gives the permissible shear stress on the vegetation. The value of C_f is found in Table 4.5.

$$\tau_p = \frac{\tau_{p,soil}}{(1 - C_f)}\left(\frac{n}{n_s}\right)^2 = \frac{0.4}{(1 - 0.9)}\left(\frac{0.032}{0.026}\right)^2 = 6.06 \text{ lb / ft}^2$$

The safety factor for this channel is taken as 1.0.

Step 7. The grass lining reinforced with the gravel/soil mixture is acceptable since the permissible shear is greater than the maximum shear.

This page intentionally left blank.

CHAPTER 5: MANUFACTURED (RECP) LINING DESIGN

Manufacturers have developed a variety of rolled erosion control products (RECPs) for erosion protection of channels. These products consist of materials that are stitched or bound into a fabric. Table 5.1 summarizes the range of RECP linings that are available from the erosion control industry. Selection of a particular product depends on the overall performance requirements for the design. RECPs offer ease of construction in climate regions where vegetation establishes quickly.

Table 5.1. Manufactured (RECP) Linings

Type	Description
Open-Weave Textile	A temporary degradable RECP composed of processed natural or polymer yarns woven into a matrix, used to provide erosion control and facilitate vegetation establishment. Examples: jute net, woven paper net, straw with net.
Erosion Control Blanket	A temporary degradable RECP composed of processed natural or polymer fibers mechanically, structurally or chemically bound together to form a continuous matrix to provide erosion control and facilitate vegetation establishment. Example: curled wood mat.
Turf-reinforcement Mat (TRM)	A non-degradable RECP composed of UV stabilized synthetic fibers, filaments, netting and/or wire mesh processed into a three-dimensional matrix. TRMs provide sufficient thickness, strength and void space to permit soil filling and establishment of grass roots within the matrix. Example: synthetic mat.

5.1 RECP PROPERTIES

The density, stiffness and thickness of light-weight manufactured linings known as rolled erosion control products (RECPs) are the main properties that relate to flow resistance and erosion control performance. There are a series of standard tests referred to as index tests that measure these physical properties. The AASHTO National Transportation Product Evaluation Program (NTPEP) (AASHTO/NTPEP, 2002) has identified a set of test methods applicable to RECPs. *Research on RECPs has not resulted in a relationship between these index tests and hydraulic properties. Hydraulic properties must be determined by full scale testing in laboratory flumes using defined testing protocols (ASTM D 6460).* Table 5.2 summarizes index tests that relate to the physical properties of density, stiffness and thickness.

Qualitatively, denser linings prevent soil from entering into the higher-velocity flow above the liner (Gharabaghi, et al., 2002; Cotton, 1993). Linings with higher tensile strength and flexural rigidity have less deformation due to shear and uplift forces of the flow and remain in closer contact with the soil. Linings with more thickness have a larger moment of inertia, which further reduces the deformation of the lining.

NTPEP also includes two bench tests developed by the Erosion Control Technology Council (ECTC) that relate to channel erosion. Table 5.3 briefly describes the bench scale test methods applicable to RECP channel linings. *The values generated from bench-scale tests are intended for qualitative comparison of products and product quality verification. These values should not be used to design a channel lining.* Because of their small scale, these tests do not reflect larger scale currents that are generated in full scale testing in laboratory flumes using defined testing protocols (AASHTO/NTPEP, 2002; Robeson, et al., 2003).

Table 5.2. Index Tests for RECPs

Property	Index Test	Description
Density	ASTM D 6475	Standard Test Method for Mass per Unit Area for Erosion Control Blankets
	ASTM D 6566	Standard Test Method for Measuring Mass per Unit Area of Turf Reinforcement Mats
	ASTM D 6567	Standard Test Method for Measuring the Light Penetration of Turf Reinforcement Mats
Stiffness	ASTM D 4595	Test Method for Tensile Properties of Geotextile by the Wide-Width Strip Method
Thickness	ASTM D 6525	Standard Test Method for Measuring Nominal Thickness of Erosion Control Products

Table 5.3. Bench-Scale Tests for RECPs

Bench Test	Description
ECTC – Draft Method 3 Channel Erosion	Standard test method for determination of RECP ability to protect soil from hydraulically induced shear stresses under bench-scale conditions.
ECTC – Draft Method 4 Germination and Plant Growth	Standard test method for determination of RECP performance in encouraging seed germination and plant growth.

Proper installation of RECPs is critical to their performance. This includes the stapling of the lining to the channel perimeter, the lapping of adjacent fabric edges and the frequency of cutoff trenches. Each manufacturer provides guidelines on installation, which should be reviewed and incorporated into installation specifications. Construction inspection should verify that all installation specifications have been met prior to acceptance.

5.2   MANNING'S ROUGHNESS

There is no single n value formula for RECPs. The roughness of these linings must be determined by full-scale testing in laboratory flumes using defined testing protocols. As with vegetated linings, the n value varies significantly with the applied shear due to the displacement of the lining by shear and uplift forces.

The designer will need to obtain from the RECP manufacturer a table of n value versus applied shear. Three n values, with the corresponding applied shear values need to be provided by the manufacturer as shown in Table 5.4. The upper shear stress should equal or exceed the lining shear, τ_l. The upper and lower shear stress values must equal twice and one-half of the middle value, respectively.

Table 5.4. Standard n value versus Applied Shear

Applied Shear, N/m^2 (lb/ft^2)	n value
$\tau_{lower} = \tau_{mid} / 2$	n_{lower}
τ_{mid}	n_{mid}
$\tau_{upper} = 2\,\tau_{mid}$	n_{upper}

This information is used to determine the following n value relationship:

$$n = a\tau_o^b \tag{5.1}$$

where,

n	=	Manning's roughness value for the specific RECP
a	=	coefficient based on Equation 5.2
b	=	exponent based on Equation 5.3
τ_o	=	mean boundary shear stress, N/m^2 (lb/ft^2)

The coefficient "a" is based on the n value at the mid-range of applied shear.

$$a = \frac{n_{mid}}{\tau_{mid}^b} \tag{5.2}$$

The exponent "b" is computed by the following equation:

$$b = -\frac{\sqrt{\ln\left(\dfrac{n_{mid}}{n_{lower}}\right)\ln\left(\dfrac{n_{upper}}{n_{mid}}\right)}}{0.693} \tag{5.3}$$

Note that exponent "b" should be a negative value.

5.3 PERMISSIBLE SHEAR STRESS

The permissible shear stress of an RECP lining is determined both by the underlying soil properties as well as those of the RECP. In the case of TRMs, the presence of vegetation also influences erosion resistance properties.

5.3.1 Effective Shear Stress

RECPs dissipate shear stress before it reaches the soil surface. When the shear stress at the soil surface is less than the permissible shear for the soil surface, then erosion of the soil surface will be controlled. RECPs provide shear reduction primarily by providing cover for the soil surface. As the hydraulic forces on the RECP lining increase, the lining is detached from the soil, which permits a current to establish between the lining and the soil surface. Turbulent fluctuations within this current eventually erode the soil surface. This process model for RECP shear on the soil surface is given by the following (See Appendix F for derivation):

$$\tau_e = \left(\tau_d - \frac{\tau_l}{4.3}\right)\left(\frac{\alpha}{\tau_l}\right) \tag{5.4}$$

where,

τ_e = effective shear stress on the soil, N/m² (lb/ft²)

τ_d = design shear stress, N/m² (lb/ft²)

τ_l = shear stress on the RECP that results in 12.5 mm (0.5 in) of erosion

α = unit conversion constant, 6.5 (SI), 0.14 (CU)

The value of τ_l is determined based on a standard soil specified in the testing protocol. Permissible shear stress for the underlying soil has been presented in Section 4.3.2. The reader is referred to that section for that discussion.

5.3.2 Permissible RECP/Soil Shear Stress

The combined effects of the soil permissible shear stress and the effective shear stress transferred through the RECP lining results in a permissible shear stress for the RECP lining. Taking Equation 5.4 and substituting the permissible shear stress for the soil for the effective shear stress on the soil, τ_e, gives the following equation for permissible shear stress for the RECP lining:

$$\tau_p = \frac{\tau_l}{\alpha}\left(\tau_{p,soil} + \frac{\alpha}{4.3}\right) \tag{5.5}$$

where,

τ_p = permissible shear stress on the RECP lining, N/m² (lb/ft²)

τ_l = shear stress on the RECP that results in 12.5 mm (0.5 in) of erosion

$\tau_{p,soil}$ = permissible soil shear stress, N/m² (lb/ft²)

α = unit conversion constant, 6.5 (SI), 0.14 (CU)

Design Example: Manufactured Lining Design (SI)

Evaluate a temporary channel lining for a roadside channel. Two alternative RECPs are available. Alternative A costs less.

Given:

Shape: Trapezoidal, B = 0.9 m, Z = 3

Soil: Clayey sand (SC classification), PI = 16, e = 0.5

Grade: 3.0 percent

Flow: 0.30 m³/s

RECP Product A:

Erosion Control Blanket, ECB, Manufacturers performance data

$\tau_l = 60$ N/m² (Shear on lining at 12.5 mm soil loss)

Roughness rating:

Applied Shear, N/m²	n value
35	0.038
70	0.034
140	0.031

RECP Product B:

Erosion Control Blanket, ECB, Manufacturers performance data

$\tau_l = 100$ N/m² (Shear on lining at 12.5 mm soil loss)

Roughness rating:

Applied Shear, N/m²	n value
50	0.040
100	0.036
200	0.033

Solution

First, try the less expensive **"Product A."** The solution is accomplished using procedure given in Section 3.1 for a straight channel.

Step 1. Channel slope, shape, and discharge have been given.

Step 2. Select erosion ECB A.

Step 3. Initial depth is estimated at 0.30 m

From the geometric relationship of a trapezoid (see Appendix B):

$$A = Bd + Zd^2 = 0.9(0.3) + 3(0.3)^2 = 0.540 \text{ m}^2$$

$$P = B + 2d\sqrt{Z^2 + 1} = 0.9 + 2(0.3)\sqrt{3^2 + 1} = 2.80 \text{ m}$$

R = A/P = (0.540)/(2.80) = 0.193 m

Step 4. To estimate n, the applied shear stress on the lining is given by Equation 2.3

$$\tau_o = \gamma R S_o = 9810(0.193)(0.03) = 56.8 \text{ N/m}^2$$

Determine a Manning's n value from Equation 5.1 with support from Equations 5.2 and 5.3.

$$b = -\frac{\sqrt{\ln\left(\frac{n_{mid}}{n_{lower}}\right)\ln\left(\frac{n_{upper}}{n_{mid}}\right)}}{0.693} = -\frac{\sqrt{\ln\left(\frac{0.034}{0.038}\right)\ln\left(\frac{0.031}{0.034}\right)}}{0.693} = -0.146$$

$$a = \frac{n_{mid}}{\tau_{mid}^b} = \frac{0.034}{70^{-0.146}} = 0.0632$$

$$n = a\tau_o^b = 0.0632(56.8)^{-0.146} = 0.035$$

The discharge is calculated using Manning's equation (Equation 2.1):

$$Q = \frac{\alpha}{n}AR^{2/3}S_f^{1/2} = \frac{1}{0.035}(0.540)(0.193)^{2/3}(0.03)^{1/2} = 0.88\ m^3/s$$

Step 5. Since this value is more than 5 percent different from the design flow, we need to go back to step 3 to estimate a new flow depth.

Step 3 (2nd iteration). Estimate a new depth solving Equation 2.2 or other appropriate method iteratively to find the next estimate for depth:

d = 0.18 m

Revised hydraulic radius.

$$A = Bd + Zd^2 = 0.9(0.18) + 3(0.18)^2 = 0.259\ m^2$$

$$P = B + 2d\sqrt{Z^2+1} = 0.9 + 2(0.18)\sqrt{3^2+1} = 2.04\ m$$

R = A/P = (0.259)/(2.04) = 0.127 m

Step 4 (2nd iteration). To estimate n, the applied shear stress on the lining is given by Equation 2.3

$$\tau_o = \gamma RS_o = 9810(0.127)(0.03) = 37.3\ N/m^2$$

Determine a Manning's n value from Equation 5.1. The exponent, b, and the coefficient, a, are unchanged from the earlier calculation.

$$n = a\tau_o^b = 0.0632(37.3)^{-0.146} = 0.037$$

The discharge is calculated using Manning's equation (Equation 2.1):

$$Q = \frac{\alpha}{n}AR^{2/3}S_f^{1/2} = \frac{1}{0.037}(0.259)(0.127)^{2/3}(0.03)^{1/2} = 0.31\ m^3/s$$

Step 5 (2nd iteration). Since this value is within 5 percent of the design flow, we can proceed to step 6.

Step 6. The maximum shear on the lining of the channel bottom is.

$$\tau_d = \gamma dS_o = 9810(0.18)(0.03) = 52.9\ N/m^2$$

Determine the permissible soil shear stress from Equation 4.6 and Table 4.6.

$$\tau_{p,soil} = \left(c_1PI^2 + c_2PI + c_3\right)\left(c_4 + c_5e\right)^2c_6 = \left(1.07(16)^2 + 14.3(16) + 47.7\right)\left(1.42 - 0.61(0.5)\right)^2(0.0048) = 3.28\ N/m^2$$

Equation 5.5 gives the permissible shear on the RECP.

$$\tau_p = \frac{\tau_l}{\alpha}\left(\tau_{p,soil} + \frac{\alpha}{4.3}\right) = \frac{60}{6.5}\left(3.28 + \frac{6.5}{4.3}\right) = 44.2\ N/m^2$$

Safety factor for this channel is selected to be equal to 1.0.

Step 7. Product A (ECB lining) is not acceptable since the maximum shear on the RECP surface is greater than the permissible shear of the RECP.

Now try the alternative **"Product B."** The flow and channel configuration as well as the permissible shear stress are the same. Also, it is reasonable to assume an initial depth equal to the last depth we calculated for Product A. Therefore, using the area and hydraulic radius from that calculation, we can start with Step 4.

Step 4. To estimate n, the applied shear stress on the lining is given by Equation 2.3

$$\tau_o = \gamma\,RS_o = 9810(0.127)(0.03) = 37.4\ N/m^2$$

Determine a Manning's n value from Equation 5.1 with support from Equations 5.2 and 5.3.

$$b = -\frac{\sqrt{\ln\left(\frac{n_{mid}}{n_{lower}}\right)\ln\left(\frac{n_{upper}}{n_{mid}}\right)}}{0.693} = -\frac{\sqrt{\ln\left(\frac{0.036}{0.040}\right)\ln\left(\frac{0.033}{0.036}\right)}}{0.693} = -0.138$$

$$a = \frac{n_{mid}}{\tau_{mid}^b} = \frac{0.036}{100^{-0.138}} = 0.0680$$

$$n = a\tau_o^b = 0.0680(37.4)^{-0.138} = 0.041$$

The discharge is calculated using Manning's equation (Equation 2.1):

$$Q = \frac{\alpha}{n}AR^{2/3}S_f^{1/2} = \frac{1}{0.041}(0.259)(0.127)^{2/3}(0.03)^{1/2} = 0.28\ m^3/s$$

Step 5 Since this value is more than 5 percent different from the design flow, we need to go back to step 3 to estimate a new flow depth.

Step 3 (2nd iteration). Estimate a new depth solving Equation 2.2 or other appropriate method iteratively to find the next estimate for depth:

d = 0.19 m

Revise hydraulic radius.

$$A = Bd + Zd^2 = 0.9(0.19) + 3(0.19)^2 = 0.279\ m^2$$

$$P = B + 2d\sqrt{Z^2 + 1} = 0.9 + 2(0.19)\sqrt{3^2 + 1} = 2.10\ m$$

R = A/P = (0.279)/(2.10) = 0.132 m

Step 4 (2nd iteration). To estimate n, the applied shear stress on the lining is given by Equation 2.3

$$\tau_o = \gamma\,RS_o = 9810(0.132)(0.03) = 38.8\ N/m^2$$

Determine a Manning's n value from Equation 5.1. The exponent, b, and the coefficient, a, are unchanged from the earlier calculation.

$$n = a\tau_o^b = 0.0680(38.8)^{-0.138} = 0.041$$

The discharge is calculated using Manning's equation (Equation 2.1):

$$Q = \frac{\alpha}{n} AR^{\frac{2}{3}} S_f^{\frac{1}{2}} = \frac{1}{0.041}(0.279)(0.132)^{\frac{2}{3}}(0.03)^{\frac{1}{2}} = 0.305 \text{ m}^3/s$$

Step 5 (2nd iteration). Since this value is within 5 percent of the design flow, we can proceed to step 6.

Step 6. The maximum shear on the lining of the channel bottom is.

$$\tau_d = \gamma dS_o = 9810(0.19)(0.03) = 55.9 \text{ N/m}^2$$

Equation 5.5 gives the permissible shear on the RECP.

$$\tau_p = \frac{\tau_l}{\alpha}\left(\tau_{p,soil} + \frac{\alpha}{4.3}\right) = \frac{100}{6.5}\left(3.28 + \frac{6.5}{4.3}\right) = 73.7 \text{ N/m}^2$$

Step 8. Product B (ECB lining) is an acceptable temporary lining since the maximum shear on the RECP surface is less than the permissible shear of the RECP. Choose Product B. (Remember the permanent vegetative lining must also be evaluated.)

Design Example: Manufactured Lining Design (CU)

Evaluate a temporary channel lining for a roadside channel. Two alternative RECPs are available. Alternative A costs less.

Given:

Shape: Trapezoidal, B = 3.0 ft, Z = 3
Soil: Clayey sand (SC classification), PI = 16, e = 0.5
Grade: 3.0 percent
Flow: 10 ft³/s

RECP Product A:

Erosion Control Blanket, ECB, Manufacturers performance data

τ_l = 1.25 lb/ft² (Shear on lining at 0.5 in soil loss)

Roughness rating:

Applied Shear, lb/ft²	n value
0.75	0.038
1.5	0.034
3.0	0.031

RECP Product B:

Erosion Control Blanket, ECB, Manufacturers performance data

$\tau_l = 2.0$ lb/ft^2 (Shear on lining at 0.5 in soil loss)

Roughness rating:

Applied Shear, lb/ft^2	n value
1.0	0.040
2.0	0.036
4.0	0.033

Solution

First, try the less expensive **"Product A."** The solution is accomplished using procedure given in Section 3.1 for a straight channel.

Step 1. Channel slope, shape, and discharge have been given.

Step 2. Try ECB Product A.

Step 3. Initial depth is estimated at 1.0 ft

From the geometric relationship of a trapezoid (see Appendix B):

$$A = Bd + Zd^2 = 3.0(1.0) + 3(1.0)^2 = 6.00 \text{ ft}^2$$

$$P = B + 2d\sqrt{Z^2 + 1} = 3.0 + 2(1.0)\sqrt{3^2 + 1} = 9.32 \text{ ft}$$

$$R = A/P = (6.00)/(9.32) = 0.644 \text{ ft}$$

Step 4. To estimate n, the applied shear stress on the lining is given by Equation 2.3

$$\tau_o = \gamma\, RS_o = 62.4(0.644)(0.03) = 1.21 \text{ lb / ft}^2$$

Determine a Manning's n value from Equation 5.1 with support from Equations 5.2 and 5.3.

$$b = -\frac{\sqrt{\ln\left(\dfrac{n_{mid}}{n_{lower}}\right)\ln\left(\dfrac{n_{upper}}{n_{mid}}\right)}}{0.693} = -\frac{\sqrt{\ln\left(\dfrac{0.034}{0.038}\right)\ln\left(\dfrac{0.031}{0.034}\right)}}{0.693} = -0.146$$

$$a = \frac{n_{mid}}{\tau_{mid}^b} = \frac{0.034}{1.5^{-0.146}} = 0.0361$$

$$n = a\tau_o^b = 0.0361(1.21)^{-0.146} = 0.035$$

The discharge is calculated using Manning's equation (Equation 2.1):

$$Q = \frac{\alpha}{n}AR^{2/3}S_f^{1/2} = \frac{1.49}{0.035}(6.0)(0.644)^{2/3}(0.03)^{1/2} = 33.0 \text{ ft}^3 / s$$

Step 5. Since this value is more than 5 percent different from the design flow, we need to go back to step 3 to estimate a new flow depth.

Step 3 (2nd iteration). Estimate a new depth solving Equation 2.2 or other appropriate method iteratively to find the next estimate for depth:

d = 0.57 ft

Revise hydraulic radius.

$$A = Bd + Zd^2 = 3.0(0.57) + 3(0.57)^2 = 2.68 \text{ ft}^2$$

$$P = B + 2d\sqrt{Z^2 + 1} = 3.0 + 2(0.57)\sqrt{3^2 + 1} = 6.60 \text{ ft}$$

$$R = A/P = (2.68)/(6.60) = 0.406 \text{ ft}$$

Step 4 (2nd iteration). To estimate n, the applied shear stress on the lining is given by Equation 2.3

$$\tau_o = \gamma RS_o = 62.4(0.406)(0.03) = 0.76 \text{ lb} / \text{ft}^2$$

Determine a Manning's n value from Equation 5.1. The exponent, b, and the coefficient, a, are unchanged from the earlier calculation.

$$n = a\tau_o^b = 0.0361(0.76)^{-0.146} = 0.037$$

The discharge is calculated using Manning's equation (Equation 2.1):

$$Q = \frac{\alpha}{n} AR^{2/3}S_f^{1/2} = \frac{1.49}{0.037}(2.68)(0.406)^{2/3}(0.03)^{1/2} = 10.2 \text{ ft}^3 / \text{s}$$

Step 5 (2nd iteration). Since this value is within 5 percent of the design flow, we can proceed to step 6.

Step 6. The maximum shear on the channel bottom is:

$$\tau_d = \gamma dS_o = 62.4(0.57)(0.03) = 1.07 \text{ lb} / \text{ft}^2$$

Determine the permissible soil shear stress from Equation 4.6 and Table 4.6.

$$\tau_{p,soil} = (c_1 PI^2 + c_2 PI + c_3)(c_4 + c_5 e)^2 c_6 = (1.07(16)^2 + 14.3(16) + 47.7)(1.42 - 0.61(0.5))^2(0.0001) = 0.068 \text{ lb}/\text{ft}^2$$

Equation 5.5 gives the permissible shear on the RECP.

$$\tau_p = \frac{\tau_l}{\alpha}\left(\tau_{p,soil} + \frac{\alpha}{4.3}\right) = \frac{1.25}{0.14}\left(0.068 + \frac{0.14}{4.3}\right) = 0.90 \text{ lb} / \text{ft}^2$$

Safety factor for this channel is selected to be equal to 1.0.

Step 7. Product A (ECB lining) is not acceptable since the maximum shear on the RECP surface is greater than the permissible shear of the RECP.

Now try the alternative **"Product B."** The flow and channel configuration as well as the permissible shear stress are the same. Also, it is reasonable to assume an initial depth equal to the last depth we calculated for Product A. Therefore, using the area and hydraulic radius from that calculation, we can start with Step 4.

Step 4. To estimate n, the applied shear stress on the lining is given by Equation 2.3

$$\tau_o = \gamma RS_o = 62.4(0.406)(0.03) = 0.76 \text{ lb} / \text{ft}^2$$

Determine a Manning's n value from Equation 5.1 with support from Equations 5.2 and 5.3.

$$b = -\frac{\sqrt{\ln\left(\dfrac{n_{mid}}{n_{lower}}\right)\ln\left(\dfrac{n_{upper}}{n_{mid}}\right)}}{0.693} = -\frac{\sqrt{\ln\left(\dfrac{0.036}{0.040}\right)\ln\left(\dfrac{0.033}{0.036}\right)}}{0.693} = -0.138$$

$$a = \frac{n_{mid}}{\tau_{mid}^b} = \frac{0.036}{2.0^{-0.138}} = 0.0396$$

$$n = a\tau_o^b = 0.0396(0.76)^{-0.138} = 0.041$$

The discharge is calculated using Manning's equation (Equation 2.1):

$$Q = \frac{\alpha}{n}AR^{\frac{2}{3}}S_f^{\frac{1}{2}} = \frac{1.49}{0.041}(2.68)(0.406)^{\frac{2}{3}}(0.03)^{\frac{1}{2}} = 9.25 \text{ ft}^3/s$$

Step 5. Since this value is more than 5 percent different from the design flow, we need to go back to step 3 to estimate a new flow depth.

Step 3 (2nd iteration). Estimate a new depth solving Equation 2.2 or other appropriate method iteratively to find the next estimate for depth:

d = 0.59 ft

Revise hydraulic radius.

$$A = Bd + Zd^2 = 3.0(0.59) + 3(0.59)^2 = 2.81 \text{ ft}^2$$

$$P = B + 2d\sqrt{Z^2 + 1} = 3.0 + 2(0.59)\sqrt{3^2 + 1} = 6.73 \text{ ft}$$

$$R = A/P = (2.81)/(6.73) = 0.418 \text{ ft}$$

Step 4 (2nd iteration). To estimate n, the applied shear stress on the lining is given by Equation 2.3

$$\tau_o = \gamma RS_o = 62.4(0.418)(0.03) = 0.78 \text{ lb}/\text{ft}^2$$

Determine a Manning's n value from Equation 5.1. The exponent, b, and the coefficient, a, are unchanged from the earlier calculation.

$$n = a\tau_o^b = 0.0396(0.78)^{-0.138} = 0.041$$

The discharge is calculated using Manning's equation (Equation 2.1):

$$Q = \frac{\alpha}{n}AR^{\frac{2}{3}}S_f^{\frac{1}{2}} = \frac{1.49}{0.041}(2.81)(0.418)^{\frac{2}{3}}(0.03)^{\frac{1}{2}} = 9.89 \text{ ft}^3/s$$

Step 5 (2nd iteration). Since this value is within 5 percent of the design flow, we can proceed to step 6.

Step 6. The maximum shear on the lining of the channel bottom is.

$$\tau_d = \gamma dS_o = 62.4(0.59)(0.03) = 1.10 \text{ lb}/\text{ft}^2$$

Equation 5.5 gives the permissible shear on the RECP.

$$\tau_p = \frac{\tau_l}{\alpha}\left(\tau_{p,soil} + \frac{\alpha}{4.3}\right) = \frac{2.0}{0.14}\left(0.068 + \frac{0.14}{4.3}\right) = 1.44 \text{ lb / ft}^2$$

Step 7. Product B (ECB lining) is an acceptable temporary lining since the maximum shear on the RECP is less than the permissible shear of the RECP. Choose Product B. (Remember the permanent vegetative lining must also be evaluated.)

5.4 TURF REINFORCEMENT WITH RECPS

Turf reinforcement integrates soil, lining material and grass/stems roots within a single matrix (Santha and Santha, 1995). Since turf reinforcement is a long-term solution, the lining consists of non-degradable materials. Turf Reinforcement Mats (TRMs), a subset of RECPs, are integrated with soil, and subsequently vegetation, by either covering the mat with soil or through surface application (no soil filling) allowing the vegetation to grow up through the TRM.

In the initial unvegetated state, the linings respond according to Equation 5.4. However, stability of the TRMs is achieved through proper installation per the manufacturer's recommended methods and use of proper length and quantity of fasteners.

As grass roots/stems develop within or through the TRM matrix, the lining becomes more integrated with the vegetation and soil. In the case of TRM linings, the plant roots/stems bind the mat, which prevents the detachment of the mat from the soil surface – significantly reducing the formation of under currents between the mat and soil. Grass growth further deflects turbulence away from the soil surface, establishing a positive relationship between lining and grass growth.

Where lining material is placed on top of the seed bed, the plant stem will grow through the lining (Lancaster, 1996). In this type of placement, the lining material offers more initial protection for the seed bed and provides stem reinforcement of the vegetation.   However, the grass stem may be less effective at securing the lining to the soil surface than the plant roots, permitting the lining to be displaced by hydraulic forces. Specific system performance is determined by the interaction of the vegetation and the soil-filled or surface applied TRM.

When lining material remains in place for the long-term, roadside maintenance activities need to be considered, particularly mowing. Use of proper installation methods, including a sufficient quantity and size of fasteners is necessary to prevent potential problems with mowing and other vegetation maintenance equipment. Additionally, proper maintenance techniques are required to avoid damage to the installation and ensure the integrity of the system over time.

5.4.1 Testing Data and Protocols

Unlike other RECPs there is no broadly accepted protocol for the testing of TRMs. This places an additional burden on the designer of a TRM lining to review and understand how each manufacturer has tested its products. The following checklist (Table 5.5) is based on ASTM D 6460 with the addition of requirements for TRM testing (Lipscomb, et al., 2003). It can be used as minimum standard to evaluate manufacturer's testing protocols. Products that are based on a testing protocol for TRMs that do not meet these minimum elements should only be used cautiously.

Test data consisting of the following is recommended:

  1. Permissible lining shear stress for a vegetated lining alone,

  2. Permissible lining shear stress for a vegetated lining with turf reinforcement.

3. A plant density factor (fractional increase or decrease in plant density) that is attributable to the lining material as a fraction of the cover.

Items (1) and (2) will vary depending on the soil and vegetation used in the testing. However, this is mitigated by their use in a relative, not absolute, manner as will be described in the next section. In addition, the definition of instability provided in Table 5.5 is qualitative and may be expected to vary from researcher to researcher. As long as a researcher maintains consistency within a set of tests for a particular TRM, the results should be acceptable.

5.4.2 Turf-Reinforcement Mat Cover Factor

A TRM modifies the cover factor for vegetated linings (Equation 4.3). The adjusted cover factor is determined by the following equation.

$$C_{f,TRM} = 1 - \left(\frac{\tau_{p,VEG-test}}{\tau_{p,TRM-test}} \right) \left(1 - C_{f,VEG} \right) \tag{5.6}$$

where,

$\tau_{p,VEG\text{-}test}$ = permissible shear stress on the vegetative lining, N/m^2 (lb/ft^2) as reported by Manufacturer's testing

$\tau_{p,TRM\text{-}test}$ = permissible shear stress on the turf-reinforced vegetative lining, N/m^2 (lb/ft^2) as reported by manufacturer's testing

$C_{f,VEG}$ = grass cover factor (see Table 4.5)

$C_{f,TRM}$ = TRM cover factor

If the manufacturer notes that the TRM affects plant cover density, this information should be used in the selection of $C_{f,veg}$ from Table 4.5.

Table 5.5. TRM Protocol Checklist

Protocol Element	Description	Check
Test Channel Preparation	In accordance with ASTM D 6460, except that soil type may vary. 1. Soil should be a suitable medium for plant material. 2. Soil should be of the same type and properties (plasticity index and D_{75}) for the test on vegetation alone and for test on vegetation with turf reinforcement.	
Calibration	In accordance with ASTM D 6460.	
Pre-Test Documentation	In accordance with ASTM D 6460. In addition, the vegetation type and density should be determined. 1. The type of plant material and habit (sod, bunched, mixed) should be identified. 2. A vegetation stem density count should be performed using a minimum 58 sq. cm. (9 sq. in.) frame at the beginning of each test.	
Test Setup	In accordance with ASTM D 6460. In addition: 1. Vegetation should be grown from seed for a minimum period of one year. 2. Prior to testing, the vegetation should be mowed to a standard height not to exceed 0.20 meters (8 inches).	
Test Operation and Data Collection	In accordance with ASTM D 6460, except that the test should not be conducted to catastrophic failure. 1. The channel surface should be inspected after hydraulic conditions have been maintained for no less than one-hour. 2. A vegetation stem density count should be performed and bed elevations measured. 3. The channel surface should then be inspected to determine the stability of the system. Instability is defined as: **"Loss of vegetation sufficient to expose the roots and subject the underlying soil to significant erosion."** 4. Upon inspection, if instability of the channel surface is observed then testing should be terminated. If instability is not noted, testing should be continued with the next target discharge.	

Design Example: Turf Reinforcement with a Turf Reinforcement Mat (SI)

Evaluate the following proposed lining design for a vegetated channel reinforced with turf reinforcement mat (TRM). The TRM will be placed into the soil and secured to channel boundary following manufacturer's recommendations. The permissible shear stress values for the TRM were developed from testing that meets the minimum requirements of Table 5.5.

Given:

    Shape: Trapezoidal, B = 0.6 m, Z = 3

    Soil: Silty sand (SM classification), PI = 17, e = 0.6

    Grass: Sod, good condition, h = 0.150 m

    Grade: 10.0 percent

    Flow: 0.25 m³/s

TRM Product Information from manufacturer:

$\tau_{p,TRM\text{-test}}$ = 550 N/m³

$\tau_{p,VEG\text{-test}}$ = 425 N/m³

Effect on plant density is negligible.

Solution

First, we will check to see if the channel is stable with a grass lining alone.

The solution is accomplished using procedure given in Section 3.1 for a straight channel.

    Step 1. Channel slope, shape, and discharge have been given.

    Step 2. A vegetative lining on silty sand soil will be evaluated

    Step 3. Initial depth is estimated at 0.30 m

              From the geometric relationship of a trapezoid (see Appendix B):

$$A = Bd + Zd^2 = 0.6(0.3) + 3(0.3)^2 = 0.450 \text{ m}^2$$

$$P = B + 2d\sqrt{Z^2 + 1} = 0.6 + 2(0.3)\sqrt{3^2 + 1} = 2.50 \text{ m}$$

$$R = A/P = (0.540)/(2.80) = 0.180 \text{ m}$$

    Step 4. Estimate the Manning's n value appropriate for the lining type from Equation 4.2, first calculating the mean boundary shear.

$$\tau_o = \gamma R S_o = 9810(0.180)(0.10) = 177 \text{ N/m}^2$$

$$n = \alpha C_n \tau_o^{-0.4} = 1.0(0.205)(177)^{-0.4} = 0.026$$

              The discharge is calculated using Manning's equation (Equation 2.1):

$$Q = \frac{\alpha}{n} AR^{2/3}S_f^{1/2} = \frac{1}{0.026}(0.450)(0.180)^{2/3}(0.10)^{1/2} = 1.8 \text{ m}^3/s$$

    Step 5. Since this value is more than 5 percent different from the design flow, we need to go back to step 3 to estimate a new flow depth.

Step 3 (2nd iteration). Estimate a new depth solving Equation 2.2 or other appropriate method iteratively to find the next estimate for depth:

d = 0.13 m

Revise the hydraulic radius

$$A = Bd + Zd^2 = 0.6(0.13) + 3(0.13)^2 = 0.129 \text{ m}^2$$

$$P = B + 2d\sqrt{Z^2 + 1} = 0.6 + 2(0.13)\sqrt{3^2 + 1} = 1.42 \text{ m}$$

R = A/P = (0.129)/(1.42) = 0.091 m

Step 4 (2nd iteration). Estimate the Manning's n value appropriate for the lining type from Equation 4.2, first calculating the mean boundary shear.

$$\tau_o = \gamma RS_o = 9810(0.091)(0.10) = 89 \text{ N/m}^2$$

Determine a Manning's n value from Equation 4.2. From Table 4.3, C_n = 0.205

$$n = \alpha C_n \tau_o^{-0.4} = 1.0(0.205)(89)^{-0.4} = 0.034$$

The discharge is calculated using Manning's equation (Equation 2.1):

$$Q = \frac{\alpha}{n} AR^{2/3}S_f^{1/2} = \frac{1}{0.034}(0.129)(0.091)^{2/3}(0.10)^{1/2} = 0.25 \text{ m}^3/\text{s}$$

Step 5 (2nd iteration). Since this value is within 5 percent of the design flow, we can proceed to step 6.

Step 6. The maximum shear on the channel bottom is.

$$\tau_d = \gamma dS_o = 9810(0.13)(0.10) = 128 \text{ N/m}^2$$

Determine the permissible soil shear stress from Equation 4.6.

$$\tau_{p,soil} = (c_1 PI^2 + c_2 PI + c_3)(c_4 + c_5 e)^2 c_6 = (1.07(17)^2 + 7.15(17) + 11.9)(1.42 - 0.61(0.6))^2(0.0048) = 2.36 \text{ N/m}^2$$

Equation 4.7 gives the permissible shear stress on the vegetation. The value of C_f is found in Table 4.5.

$$\tau_{p,VEG} = \frac{\tau_{p,soil}}{(1-C_f)}\left(\frac{n}{n_s}\right)^2 = \frac{2.36}{(1-0.90)}\left(\frac{0.034}{0.016}\right)^2 = 107 \text{ N/m}^2$$

The safety factor for this channel is taken as 1.0.

Step 7. The grass lining is not acceptable since the maximum shear on the vegetation, 128 N/m^2 is more than the permissible shear of grass lining, 107 N/m^2.

Now try the same grass lining with turf reinforcement. The flow and channel configuration are the same. Therefore, we begin at step 6.

Step 6. The maximum shear on the channel bottom and the permissible soil shear are the same as in the previous iteration. A new cover factor is computed based on the TRM properties. Equation 4.7 gives the permissible shear stress on the vegetation. The value of C_f is computed using Equation 5.6

$$C_{f,TRM} = 1 - \left(\frac{\tau_{p,VEG}}{\tau_{p,TRM}}\right)\left(1 - C_{f,VEG}\right) = 1 - \left(\frac{425}{550}\right)\left(1 - 0.90\right) = 0.923$$

$$\tau_{p,VEG} = \frac{\tau_{p,soil}}{\left(1 - C_f\right)}\left(\frac{n}{n_s}\right)^2 = \frac{2.36}{\left(1 - 0.923\right)}\left(\frac{0.034}{0.016}\right)^2 = 138\ N/m^2$$

The safety factor for this channel is taken as 1.0.

Step 7. The turf reinforced grass lining is acceptable since the maximum shear on the vegetation, 128 N/m^2 is less than the permissible shear of the reinforced grass lining, 138 N/m^2.

Design Example: Turf Reinforcement with a Turf Reinforcement Mat (CU)

Evaluate the following proposed lining design for a vegetated channel reinforced with turf reinforcement mat (TRM). The TRM will be placed into the soil and secured to channel boundary following manufacturer's recommendations. The permissible shear stress values for the TRM were developed from testing that meets the minimum requirements of Table 5.5.

Given:

Shape: Trapezoidal, B = 2.0 ft, Z = 3

Soil: Silty sand (SM classification), PI = 16, e = 0.6

Grass: Sod, good condition, h = 0.5 ft

Grade: 10.0 percent

Flow: 8.8 ft^3/s

TRM Product Information from Manufacturer:

$\tau_{p,TRM-test}$ = 11.5 lb/ft^3

$\tau_{p,VEG-test}$ = 8.9 lb/ft^3

Effect on plant density is negligible.

Solution

First, we will check to see if the channel is stable with a grass lining alone.

The solution is accomplished using procedure given in Section 3.1 for a straight channel.

Step 1. Channel slope, shape, and discharge have been given.

Step 2. A vegetative lining on silty sand soil will be evaluated

Step 3. Initial depth is estimated at 1.0 ft

From the geometric relationship of a trapezoid (see Appendix B):

$$A = Bd + Zd^2 = 2.0(1.0) + 3(1.0)^2 = 5.00\ ft^2$$

$$P = B + 2d\sqrt{Z^2 + 1} = 2.0 + 2(1.0)\sqrt{3^2 + 1} = 8.32\ ft$$

R = A/P = (5.00)/(8.32) = 0.601 ft

Step 4. Estimate the Manning's n value appropriate for the lining type from Equation 4.2, first calculating the mean boundary shear.

$$\tau_o = \gamma \, RS_o = 62.4(0.601)(0.10) = 3.75 \, \text{lb} / \text{ft}^2$$

$$n = \alpha C_n \tau_o^{-0.4} = 0.213(0.205)(3.75)^{-0.4} = 0.026$$

The discharge is calculated using Manning's equation (Equation 2.1):

$$Q = \frac{\alpha}{n} AR^{\frac{2}{3}} S_f^{\frac{1}{2}} = \frac{1.49}{0.026}(5.0)(0.601)^{\frac{2}{3}}(0.10)^{\frac{1}{2}} = 65 \, \text{ft}^3 / \text{s}$$

Step 5. Since this value is more than 5 percent different from the design flow, we need to go back to step 3 to estimate a new flow depth.

Step 3 (2nd iteration). Estimate a new depth solving Equation 2.2 or other appropriate method iteratively to find the next estimate for depth:

d = 0.43 ft

Revise the hydraulic radius

$$A = Bd + Zd^2 = 2.0(0.43) + 3(0.43)^2 = 1.41 \, \text{ft}^2$$

$$P = B + 2d\sqrt{Z^2 + 1} = 2.0 + 2(0.43)\sqrt{3^2 + 1} = 4.72 \, \text{ft}$$

R = A/P = (1.41)/(4.72) = 0.299 ft

Step 4 (2nd iteration). To estimate n, the applied shear stress on the grass lining is given by Equation 2.3

$$\tau_o = \gamma \, RS_o = 62.4(0.299)(0.10) = 1.87 \, \text{lb} / \text{ft}^2$$

$$n = \alpha C_n \tau_o^{-0.4} = 0.213(0.205)(1.87)^{-0.4} = 0.034$$

The discharge is calculated using Manning's equation (Equation 2.1):

$$Q = \frac{\alpha}{n} AR^{\frac{2}{3}} S_f^{\frac{1}{2}} = \frac{1.49}{0.034}(1.41)(0.299)^{\frac{2}{3}}(0.10)^{\frac{1}{2}} = 8.7 \, \text{ft}^3 / \text{s}$$

Step 5 (2nd iteration). Since this value is within 5 percent of the design flow, we can proceed to step 6.

Step 6. The maximum shear on the channel bottom is.

$$\tau_d = \gamma dS_o = 62.4(0.43)(0.10) = 2.7 \, lb / ft^2$$

Determine the permissible soil shear stress from Equation 4.6.

$$\tau_{p,soil} = \left(c_1 PI^2 + c_2 PI + c_3\right)\left(c_4 + c_5 e\right)^2 c_6 = \left(1.07(17)^2 + 7.15(17) + 11.9\right)\left(1.42 - 0.61(0.6)\right)^2 (0.0001) = 0.049 \, \text{lb} / \text{ft}^2$$

Equation 4.7 gives the permissible shear stress on the vegetation. The value of C_f is found in Table 4.5.

$$\tau_{p,VEG} = \frac{\tau_{p,soil}}{(1-C_f)}\left(\frac{n}{n_s}\right)^2 = \frac{0.049}{(1-0.90)}\left(\frac{0.034}{0.016}\right)^2 = 2.2 \text{ lb} / \text{ft}^2$$

The safety factor for this channel is taken as 1.0.

Step 7. The grass lining is not acceptable since the maximum shear on the vegetation, 2.7 lb/ft^2 is more than the permissible shear of grass lining, 2.2 lb/ft^2.

Now try the same grass lining with turf reinforcement. The flow and channel configuration are the same. Therefore, we begin at step 6.

Step 6. The maximum shear on the channel bottom and the permissible soil shear are the same as in the previous iteration. A new cover factor is computed based on the TRM properties. Equation 4.7 gives the permissible shear stress on the vegetation. The value of C_f is found in Table 4.5.

$$C_{f,TRM} = 1 - \left(\frac{\tau_{p,VEG}}{\tau_{p,TRM}}\right)(1 - C_{f,VEG}) = 1 - \left(\frac{8.9}{11.5}\right)(1 - 0.90) = 0.923$$

$$\tau_{p,VEG} = \frac{\tau_{p,soil}}{(1-C_f)}\left(\frac{n}{n_s}\right)^2 = \frac{0.049}{(1-0.923)}\left(\frac{0.034}{0.016}\right)^2 = 2.9 \text{ lb} / \text{ft}^2$$

The safety factor for this channel is taken as 1.0.

Step 7. The turf reinforced grass lining is acceptable since the maximum shear on the vegetation, 2.7 lb/ft^2 is less than the permissible shear of the reinforced grass lining, 2.9 lb/ft^2.

This page intentionally left blank.

CHAPTER 6: RIPRAP, COBBLE, AND GRAVEL LINING DESIGN

Riprap, cobble, and gravel linings are considered permanent flexible linings. They may be described as a noncohesive layer of stone or rock with a characteristic size, which for the purposes of this manual is the D_{50}. The applicable sizes for the guidance in this manual range from 15 mm (0.6 in) gravel up to 550 mm (22 in) riprap. For the purposes of this manual, the boundary between gravel, cobble, and riprap sizes will be defined by the following ranges:

- Gravel: 15 - 64 mm (0.6 - 2.5 in)

- Cobble: 64 - 130 mm (2.5 - 5.0 in)

- Riprap: 130 – 550 mm (5.0 – 22.0 in)

Other differences between gravels, cobbles, and riprap may include gradation and angularity. These issues will be addressed later.

Gravel mulch, although considered permanent, is generally used as supplement to aid in the establishment of vegetation (See Chapter 4). It may be considered for areas where vegetation establishment is difficult, for example, in arid-region climates. For the transition period before the establishment of the vegetation, the stability of gravel mulch should be assessed using the procedures in this chapter.

The procedures in this chapter are applicable to uniform prismatic channels (as would be characteristic of roadside channels) with rock sizes within the range given above. For situations not satisfying these two conditions, the designer is referred to another FHWA circular (No. 11) "Design of Riprap Revetment" (FHWA, 1989).

6.1 MANNING'S ROUGHNESS

Manning's roughness is a key parameter needed for determining the relationships between depth, velocity, and slope in a channel. However, for gravel and riprap linings, roughness has been shown to be a function of a variety of factors including flow depth, D_{50}, D_{84}, and friction slope, S_f. A partial list of roughness relationships includes Blodgett (1986a), Limerinos (1970), Anderson, et al. (1970), USACE (1994), Bathurst (1985), and Jarrett (1984). For the conditions encountered in roadside and other small channels, the relationships of Blodgett and Bathurst are adopted for this manual.

Blodgett (1986a) proposed a relationship for Manning's roughness coefficient, n, that is a function of the flow depth and the relative flow depth (d_a/D_{50}) as follows:

$$n = \frac{\alpha \, d_a^{1/6}}{2.25 + 5.23 \log\left(\dfrac{d_a}{D_{50}}\right)}$$

(6.1)

where,

n = Manning's roughness coefficient, dimensionless
d_a = average flow depth in the channel, m (ft)
D_{50} = median riprap/gravel size, m (ft)
α = unit conversion constant, 0.319 (SI) and 0.262 (CU)

Equation 6.1 is applicable for the range of conditions where $1.5 \le d_a/D_{50} \le 185$. For small channel applications, relative flow depth should not exceed the upper end of this range.

Some channels may experience conditions below the lower end of this range where protrusion of individual riprap elements into the flow field significantly changes the roughness relationship. This condition may be experienced on steep channels, but also occurs on moderate slopes. The relationship described by Bathurst (1991) addresses these conditions and can be written as follows (See Appendix D for the original form of the equation):

$$n = \frac{\alpha \, d_a^{1/6}}{\sqrt{g} \; f(Fr) \; f(REG) \; f(CG)}$$

(6.2)

where,

d_a = average flow depth in the channel, m (ft)

g = acceleration due to gravity, 9.81 m/s^2 (32.2 ft/s^2)

Fr = Froude number

REG = roughness element geometry

CG = channel geometry

α = unit conversion constant, 1.0 (SI) and 1.49 (CU)

Equation 6.2 is a semi-empirical relationship applicable for the range of conditions where $0.3 < d_a/D_{50} < 8.0$. The three terms in the denominator represent functions of Froude number, roughness element geometry, and channel geometry given by the following equations:

$$f(Fr) = \left(\frac{0.28 Fr}{b} \right)^{\log(0.755/b)}$$

(6.3)

$$f(REG) = 13.434 \left(\frac{T}{D_{50}} \right)^{0.492} b^{1.025(T/D_{50})^{0.118}}$$

(6.4)

$$f(CG) = \left(\frac{T}{d_a} \right)^{-b}$$

(6.5)

where,

T = channel top width, m (ft)

b = parameter describing the effective roughness concentration.

The parameter b describes the relationship between effective roughness concentration and relative submergence of the roughness bed. This relationship is given by:

$$b = 1.14 \left(\frac{D_{50}}{T} \right)^{0.453} \left(\frac{d_a}{D_{50}} \right)^{0.814}$$

(6.6)

Equations 6.1 and 6.2 both apply in the overlapping range of $1.5 \leq d_a/D_{50} \leq 8$. For consistency and ease of application over the widest range of potential design situations, use of the Blodgett equation (6.1) is recommended when $1.5 \leq d_a/D_{50}$. The Bathurst equation (6.2) is recommended for $0.3 < d_a/D_{50} < 1.5$.

As a practical problem, both Equations 6.1 and 6.2 require depth to estimate n while n is needed to determine depth setting up an iterative process.

6.2 PERMISSIBLE SHEAR STRESS

Values for permissible shear stress for riprap and gravel linings are based on research conducted at laboratory facilities and in the field. The values presented here are judged to be conservative and appropriate for design use. Permissible shear stress is given by the following equation:

$$\tau_p = F_*(\gamma_s - \gamma)D_{50} \tag{6.7}$$

where,

τ_p = permissible shear stress, N/m^2 (lb/ft^2)

F_* = Shield's parameter, dimensionless

γ_s = specific weight of the stone, N/m^3 (lb/ft^3)

γ = specific weight of the water, 9810 N/m^3 (62.4 lb/ft^3)

D_{50} = mean riprap size, m (ft)

Typically, a specific weight of stone of 25,900 N/m^3 (165 lb/ft^3) is used, but if the available stone is different from this value, the site-specific value should be used.

Recalling Equation 3.2,

$$\tau_p \geq SF\tau_d$$

and Equation 3.1,

$$\tau_d = \gamma dS_o$$

Equation 6.7 can be written in the form of a sizing equation for D_{50} as shown below:

$$D_{50} \geq \frac{SF\, d\, S_o}{F_*(SG - 1)} \tag{6.8}$$

where,

d = maximum channel depth, m (ft)

SG = specific gravity of rock (γ_s/γ), dimensionless

Changing the inequality sign to an equality gives the minimum stable riprap size for the channel bottom. Additional evaluation for the channel side slope is given in Section 6.3.2.

Equation 6.8 is based on assumptions related to the relative importance of skin friction, form drag, and channel slope. However, skin friction and form drag have been documented to vary resulting in reports of variations in Shield's parameter by different investigators, for example Gessler (1965), Wang and Shen (1985), and Kilgore and Young (1993). This variation is usually linked to particle Reynolds number as defined below:

$$R_e = \frac{V_* D_{50}}{\nu} \tag{6.9}$$

where,

R_e = particle Reynolds number, dimensionless
V_* = shear velocity, m/s (ft/s)
ν = kinematic viscosity, 1.131×10^{-6} m²/s at 15.5 deg C (1.217×10^{-5} ft²/s at 60 deg F)

Shear velocity is defined as:

$$V_* = \sqrt{gdS} \tag{6.10}$$

where,

g = gravitational acceleration, 9.81 m/s² (32.2 ft/s²)
d = maximum channel depth, m (ft)
S = channel slope, m/m (ft/ft)

Higher Reynolds number correlates with a higher Shields parameter as is shown in Table 6.1. For many roadside channel applications, Reynolds number is less than 4×10^4 and a Shields parameter of 0.047 should be used in Equations 6.7 and 6.8. In cases for a Reynolds number greater than 2×10^5, for example, with channels on steeper slopes, a Shields parameter of 0.15 should be used. Intermediate values of Shields parameter should be interpolated based on the Reynolds number.

Table 6.1. Selection of Shields' Parameter and Safety Factor

Reynolds number	F*	SF
$\leq 4 \times 10^4$	0.047	1.0
$4 \times 10^4 < R_e < 2 \times 10^5$	Linear interpolation	Linear interpolation
$\geq 2 \times 10^5$	0.15	1.5

Higher Reynolds numbers are associated with more turbulent flow and a greater likelihood of lining failure with variations of installation quality. Because of these conditions, it is recommended that the Safety Factor be also increased with Reynolds number as shown in Table 6.1. Depending on site-specific conditions, safety factor may be further increased by the designer, but should not be decreased to values less than those in Table 6.1.

As channel slope increases, the balance of resisting, sliding, and overturning forces is altered slightly. Simons and Senturk (1977) derived a relationship that may be expressed as follows:

$$D_{50} \geq \frac{SF\, d\, S\, \Delta}{F_*(SG-1)} \tag{6.11}$$

where,

Δ = function of channel geometry and riprap size

The parameter Δ can be defined as follows (see Appendix D for the derivation):

$$\Delta = \frac{K_1(1 + \sin(\alpha + \beta))\tan\phi}{2(\cos\theta \tan\phi - SF \sin\theta \cos\beta)} \tag{6.12}$$

where,

α =   angle of the channel bottom slope

β =   angle between the weight vector and the weight/drag resultant vector in the plane of the side slope

θ =   angle of the channel side slope

ϕ =   angle of repose for the riprap

Finally, β is defined by:

$$\beta = \tan^{-1}\left(\frac{\cos\alpha}{\dfrac{2\sin\theta}{\eta\tan\phi} + \sin\alpha}\right) \tag{6.13}$$

where,

η =   stability number

The stability number is calculated using:

$$\eta = \frac{\tau_s}{F_*(\gamma_s - \gamma)D_{50}} \tag{6.14}$$

Riprap stability on a steep slope depends on forces acting on an individual stone making up the riprap. The primary forces include the average weight of the stones and the lift and drag forces induced by the flow on the stones. On a steep slope, the weight of a stone has a significant component in the direction of flow. Because of this force, a stone within the riprap will tend to move in the flow direction more easily than the same size stone on a milder gradient. As a result, for a given discharge, steep slope channels require larger stones to compensate for larger forces in the flow direction and higher shear stress.

The size of riprap linings increases quickly as discharge and channel gradient increase. Equation 6.11 is recommended when channel slope is greater than 10 percent and provides the riprap size for the channel bottom and sides. Equation 6.8 is recommended for slopes less than 5 percent. For slopes between 5 percent and 10 percent, it is recommended that both methods be applied and the larger size used for design. Values for safety factor and Shields parameter are taken from Table 6.1 for both equations.

6.3   DESIGN PROCEDURE

In this section a design procedure for riprap and gravel linings is outlined. First, the basic design procedure for selecting the riprap/gravel size for the bottom of a straight channel is given. Subsequent sections provide guidance for sizing material on the channel side slopes and adjusting for channel bends.

6.3.1 Basic Design

The riprap and gravel lining design procedure for the bottom of a straight channel is described in the following steps. It is iterative by necessity because flow depth, roughness, and shear stress are interdependent. The procedure requires the designer to specify a channel shape and slope as a starting point and is outlined in the eight-step process identified below. In this approach, the designer begins with a design discharge and calculates an acceptable D_{50} to line the channel bottom. An alternative analytical framework is to use the maximum discharge approach described in Section 3.6. For the maximum discharge approach, the designer selects the D_{50}, and determines the maximum depth and flow permitted in the channel while maintaining a stable lining. The following steps are recommended for the standard design.

Step 1. Determine channel slope, channel shape, and design discharge.

Step 2. Select a trial (initial) D_{50}, perhaps based on available sizes for the project. (Also, determine specific weight of proposed stone.)

Step 3. Estimate the depth. For the first iteration, select a channel depth, d_i. For subsequent iterations, a new depth can be estimated from the following equation or any other appropriate method.

$$d_{i+1} = d_i \left(\frac{Q}{Q_i} \right)^{0.4}$$

Determine the average flow depth, d_a in the channel. $d_a = A/T$

Step 4. Estimate Manning's n and the implied discharge. First, calculate the relative depth ratio, d_a/D_{50}. If d_a/D_{50} is greater than or equal to 1.5, use Equation 6.1 to calculate Manning's n. If d_a/D_{50} is less than 1.5 use Equation 6.2 to calculate Manning's n. Calculate the discharge using Manning's equation.

Step 5. If the calculated discharge is within 5 percent of the design discharge, then proceed to step 6. If not, go back to step 3 and estimate a new flow depth.

Step 6. Calculate the particle Reynolds number using Equation 6.6 and determine the appropriate Shields parameter and Safety Factor values from Table 6.1. If channel slope is less than 5 percent, calculate required D_{50} from Equation 6.8. If channel slope is greater than 10 percent, use Equation 6.11. If channel slope is between 5 and 10 percent, use both Equations 6.8 and 6.11 and take the largest value.

Step 7. If the D_{50} calculated is greater than the trial size in step 2, then the trial size is too small and unacceptable for design. Repeat procedure beginning at step 2 with new trial value of D_{50}. If the D_{50} calculated in step 6 is less than or equal to the previous trial size, then the previous trial size is acceptable. However, if the D_{50} calculated in step 6 is sufficiently smaller than the previous trial size, the designer may elect to repeat the design procedure at step 2 with a smaller, more cost-effective, D_{50}.

Design Example: Riprap Channel (Mild Slope) (SI)

Design a riprap lining for a trapezoidal channel. Given:

Q = 1.13 m³/s
B = 0.6 m
Z = 3
S_o = 0.02 m/m

Solution

Step 1. Channel characteristics and design discharge are given above.

Step 2. Available riprap sizes include Class 1: D_{50} = 125 mm, Class 2: D_{50} = 150 mm, Class 3: D_{50} = 250 mm. γ_s=25,900 N/m³ for all classes. Try Class 1 riprap for initial trial. D_{50}=125 mm

Step 3. Assume an initial trial depth of 0.5 m

Using the geometric properties of a trapezoid, the maximum and average flow depths are found:

$A = Bd+Zd^2 = 0.6(0.5)+3(0.5)^2 = 1.05$ m²

$$P = B + 2d\sqrt{Z^2 + 1} = 0.6 + 2(0.5)\sqrt{3^2 + 1} = 3.76 \text{ m}$$

$R = A/P = 1.05/3.76 = 0.279$ m

$T = B+2dZ = 0.6+2(0.5)(3) = 3.60$ m

$d_a = A/T = 1.05/3.60 = 0.292$ m

Step 4. The relative depth ratio, d_a/D_{50} = 0.292/0.125 = 2.3. Therefore, use Equation 6.1 to calculate Manning's n.

$$n = \frac{\alpha \, d_a^{1/6}}{2.25 + 5.23 \log\left(\dfrac{d_a}{D_{50}}\right)} = \frac{0.319 \, (0.292)^{1/6}}{2.25 + 5.23 \log\left(\dfrac{0.292}{0.125}\right)} = 0.062$$

Calculate Q using Manning's equation:

$$Q = \frac{\alpha}{n} AR^{2/3}S^{1/2} = \frac{1.0}{0.062}(1.05)(0.279)^{2/3}(0.02)^{1/2} = 1.02 \text{ m}^3/s$$

Step 5. Since this estimate is more than 5 percent from the design discharge, estimate a new depth in step 3.

Step 3 (2nd iteration). Estimate a new depth estimate:

$$d_{i+1} = d_i \left(\frac{Q}{Q_i}\right)^{0.4} = 0.5\left(\frac{1.13}{1.02}\right)^{0.4} = 0.521 \text{ m}$$

Using the geometric properties of a trapezoid, the maximum and average flow depths are found:

$A = Bd+Zd^2 = 0.6(0.521)+3(0.521)^2 = 1.13$ m²

$$P = B + 2d\sqrt{Z^2 + 1} = 0.6 + 2(0.521)\sqrt{3^2 + 1} = 3.90 \text{ m}$$

$$R = A/P = 1.13/3.90 = 0.289 \text{ m}$$

$$T = B + 2dZ = 0.6 + 2(0.521)(3) = 3.73 \text{ m}$$

$$d_a = A/T = 1.13/3.73 = 0.303 \text{ m}$$

Step 4. (2nd iteration). The relative depth ratio, $d_a/D_{50} = 0.302/0.125 = 2.4$. Therefore, use Equation 6.1 to calculate Manning's n.

$$n = \frac{\alpha\, d_a^{1/6}}{2.25 + 5.23 \log\left(\dfrac{d_a}{D_{50}}\right)} = \frac{0.319\,(0.302)^{1/6}}{2.25 + 5.23 \log\left(\dfrac{0.302}{0.125}\right)} = 0.061$$

Calculate Q using Manning's equation:

$$Q = \frac{\alpha}{n} AR^{2/3} S^{1/2} = \frac{1.0}{0.061}(1.13)(0.289)^{2/3}(0.02)^{1/2} = 1.14 \text{ m}^3/\text{s}$$

Step 5 (2nd iteration). Since this estimate is within 5 percent of the design discharge, proceed to step 6 with the most recently calculated depth.

Step 6.   Need to calculate the shear velocity and Reynolds number to determine Shields' parameter and SF.

Shear velocity, $V_* = \sqrt{gdS} = \sqrt{9.81(0.521)(0.02)} = 0.320 \text{ m}/\text{s}$

Reynolds number, $R_e = \dfrac{V_* D_{50}}{\nu} = \dfrac{0.320(0.125)}{1.131 \times 10^{-6}} = 3.5 \times 10^4$

Since $R_e \leq 4 \times 10^4$, $F_* = 0.047$ and SF = 1.0

Since channel slope is less than 5 percent, use Equation 6.8 to calculate minimum stable D_{50}.

$$SG = \gamma_s/\gamma_w = 25,900/9810 = 2.64$$

$$D_{50} \geq \frac{SF\, d\, S_o}{F_*(SG - 1)} = \frac{1.0(0.521)(0.02)}{0.047(2.64 - 1)} = 0.135 \text{ m}$$

Step 7.   The stable D_{50} is slightly larger than the Class 1 riprap, therefore Class 1 riprap is insufficient. Class 2 should be specified. The suitability of Class 2 should be verified by repeating the design procedure starting at step 2.

Design Example: Riprap Channel (Mild Slope) (CU)

Design a riprap lining for a trapezoidal channel. Given:

Q	=	40 ft^3/s
B	=	2.0 ft
Z	=	3
S$_o$	=	0.02 ft/ft

Solution

Step 1. Channel characteristics and design discharge are given above.

Step 2. Available riprap sizes include Class 1: D_{50} = 5 in, Class 2: D_{50} = 6 in, Class 3: D_{50} = 10 in. γ_s=165 lb/ft^3 for all classes. Try Class 1 riprap for initial trial. D_{50}=(5/12)=0.42 ft

Step 3. Assume an initial trial depth of 1.5 ft.

Using the geometric properties of a trapezoid, the maximum and average flow depths are found:

$A = Bd+Zd^2 = 2.0(1.5)+3(1.5)^2 = 9.75$ ft^2

$P = B + 2d\sqrt{Z^2 + 1} = 2.0 + 2(1.5)\sqrt{3^2 + 1} = 11.5$ ft

$R = A/P = 9.75/11.5 = 0.848$ ft

$T = B+2dZ = 2.0+2(1.5)(3) = 11.0$ ft

$d_a = A/T = 9.75/11.0 = 0.886$ ft

Step 4. The relative depth ratio, d_a/D_{50} = 0.886/0.42 = 2.1. Therefore, use Equation 6.1 to calculate Manning's n.

$$n = \frac{0.262\, d_a^{1/6}}{2.25 + 5.23 \log\left(\dfrac{d_a}{D_{50}}\right)} = \frac{0.262\, (0.886)^{1/6}}{2.25 + 5.23 \log\left(\dfrac{0.886}{0.42}\right)} = 0.065$$

Calculate Q using Manning's equation:

$$Q = \frac{1.49}{n} AR^{2/3}S^{1/2} = \frac{1.49}{0.065}(9.75)(0.849)^{2/3}(0.02)^{1/2} = 28.3 \text{ ft}^3/s$$

Step 5. Since this estimate is more than 5 percent from the design discharge, estimate a new depth in step 3.

Step 3 (2nd iteration). Estimate a new depth estimate:

$$d_{i+1} = d_i \left(\frac{Q}{Q_i}\right)^{0.4} = 1.5\left(\frac{40}{28.3}\right)^{0.4} = 1.72 \text{ ft}$$

Using the geometric properties of a trapezoid, the maximum and average flow depths are found:

$A = Bd+Zd^2 = 2.0(1.72)+3(1.72)^2 = 12.3$ ft^2

$P = B + 2d\sqrt{Z^2 + 1} = 2.0 + 2(1.72)\sqrt{3^2 + 1} = 12.9$ ft

$R = A/P = 12.3/12.9 = 0.953$ ft

$T = B+2dZ = 2.0+2(1.72)(3) = 12.3$ ft

$d_a = A/T = 12.3/12.3 = 1.0$ ft

Step 4. (2nd iteration). The relative depth ratio, d_a/D_{50} = 1.0/0.42 = 2.4. Therefore, use Equation 6.1 to calculate Manning's n.

$$n = \frac{0.262\, d_a^{1/6}}{2.25 + 5.23 \log\left(\dfrac{d_a}{D_{50}}\right)} = \frac{0.262\, (1.0)^{1/6}}{2.25 + 5.23 \log\left(\dfrac{1.0}{0.42}\right)} = 0.062$$

Calculate Q using Manning's equation:

$$Q = \frac{1.49}{n} AR^{2/3} S^{1/2} = \frac{1.49}{0.062} (12.3)(0.956)^{2/3} (0.02)^{1/2} = 40.6 \text{ ft}^3/s$$

Step 5 (2nd (iteration). Since this estimate is within 5 percent of the design discharge, proceed to step 6 with the most recently calculated depth.

Step 6.   Need to calculate the shear velocity and Reynolds number to determine Shields' parameter and SF.

Shear velocity, $V_* = \sqrt{gdS} = \sqrt{32.2(1.72)(0.02)} = 1.05 \text{ ft}/s$

Reynolds number, $R_e = \dfrac{V_* D_{50}}{\nu} = \dfrac{1.05(0.42)}{1.217 \times 10^{-5}} = 3.6 \times 10^4$

Since $R_e \leq 4 \times 10^4$, $F_* = 0.047$ and SF = 1.0

Since channel slope is less than 5 percent, use Equation 6.8 to calculate minimum stable D_{50}.

SG = γ_s / γ_w = 165/62.4 = 2.64

$$D_{50} \geq \frac{SF\, d\, S_o}{F_*(SG - 1)} = \frac{1.0(1.72)(0.02)}{0.047(2.64 - 1)} = 0.45 \text{ ft}$$

Step 7.   The stable D_{50} is slightly larger than the Class 1 riprap, therefore Class 1 riprap is insufficient. Class 2 should be specified. The suitability of Class 2 should be verified by repeating the design procedure starting at step 2.

6.3.2   Side Slopes

As was explained in Chapter 3, the shear stress on the channel side is less than the maximum shear stress occurring on the channel bottom as was described in Equation 3.2 repeated below.

$$\tau_s = K_1 \tau_d$$

However, since gravel and riprap linings are noncohesive, as the angle of the side slope approaches the angle of repose of the channel lining, the lining material becomes less stable. The stability of a side slope lining is a function of the channel side slope and the angle of repose of the rock/gravel lining material. This essentially results in a lower permissible shear stress on the side slope than on the channel bottom.

These two counterbalancing effects lead to the following design equation for specifying a stone size for the side slope given the stone size required for a stable channel bottom. The following equation is used if Equation 6.8 is used to size the channel bottom stone. If Equation 6.11 was used, the D_{50} from Equation 6.11 is used for the channel bottom and sides.

$$D_{50,s} = \frac{K_1}{K_2} D_{50,b} \tag{6.15}$$

where,

$D_{50,s}$ = D_{50} required for a stable side slope, m (ft)
$D_{50,b}$ = D_{50} required for a stable channel bottom, m (ft)
K_1 = ratio of channel side to bottom shear stress (see Section 3.2)
K_2 = tractive force ratio (Anderson, et al., 1970)

K_2 is a function of the side slope angle and the stone angle of repose and is determined from the following equation.

$$K_2 = \sqrt{1 - \left(\frac{\sin\theta}{\sin\phi}\right)^2}$$ (6.16)

where,

θ = angle of side slope
ϕ = angle of repose

When the side slope is represented as 1:Z (vertical to horizontal), the angle of side slope is:

$$\theta = \tan^{-1}\left(\frac{1}{Z}\right)$$ (6.17)

Angle of repose is a function of both the stone size and angularity and may be determined from Figure 6.1.

Channels lined with gravel or riprap on side slopes steeper than 1:3 may become unstable and should be avoided where feasible. If steeper side slopes are required they should be assessed using both Equation 6.11 and Equation 6.8 (in conjunction with Equation 6.15) taking the largest value for design.

Design Example: Riprap Channel Side Slope Assessment (SI)

Consider the stability of the side slopes for the design example in Section 6.3.1. Recall that a class 1 riprap (D_{50}=0.125 m) was evaluated and found to be unstable. Stable D_{50} was determined to be 0.135 m and a class 2 riprap (D_{50}=0.150 m) was recommended. Assess stability on the side slope of the trapezoidal channel.

Given:

Riprap angle of repose = 38 degrees
Z   = 3

Solution

Step 1. Calculate K_1. From Equation 3.4, K_1 is a function of Z (for Z between 1.5 and 5)

$K_1 = 0.066Z + 0.67 = 0.066(3)+0.67 = 0.87$

Step 2. Calculate K_2 from Equation 6.16 with θ calculated from Equation 6.17.

$$\theta = \tan^{-1}\left(\frac{1}{Z}\right) = \tan^{-1}\left(\frac{1}{3}\right) = 18.4 \text{ degrees}$$

$$K_2 = \sqrt{1 - \left(\frac{\sin\theta}{\sin\phi}\right)^2} = \sqrt{1 - \left(\frac{\sin(18.4)}{\sin(38)}\right)^2} = 0.86$$

Step 3.   Calculate the stable D_{50} for the side slope using Equation 6.15.

$$D_{50,s} = \frac{K_1}{K_2}D_{50,b} = \frac{0.87}{0.86}0.135 = 0.137 \text{ m}$$

Since 0.137 m is greater than the Class 1 D_{50} selected the side slope is also unstable. However, like for the channel bottom, Class 2 (D_{50}=0.150 m) would provide a stable side slope.

Design Example: Riprap Channel Side Slope Assessment (CU)

Consider the stability of the side slopes for the design example in Section 6.3.1. Recall that a class 1 riprap (D_{50}=5 in) was evaluated and found to be unstable. Stable D_{50} was determined to be 5.4 in and a class 2 riprap (D_{50}=6 in) was recommended. Assess stability on the side slope of the trapezoidal channel.

Given:

    Riprap angle of repose = 38 degrees

    Z   = 3

Solution

Step 1.   Calculate K_1. From Equation 3.4, K_1 is a function of Z (for Z between 1.5 and 5)

    K_1 = 0.066Z + 0.67 = 0.066(3)+0.67 = 0.87

Step 2.   Calculate K_2 from Equation 6.16 with θ calculated from Equation 6.17.

$$\theta = \tan^{-1}\left(\frac{1}{Z}\right) = \tan^{-1}\left(\frac{1}{3}\right) = 18.4 \text{ degrees}$$

$$K_2 = \sqrt{1 - \left(\frac{\sin\theta}{\sin\phi}\right)^2} = \sqrt{1 - \left(\frac{\sin(18.4)}{\sin(38)}\right)^2} = 0.86$$

Step 3.   Calculate the stable D_{50} for the side slope using Equation 6.15.

$$D_{50,s} = \frac{K_1}{K_2}D_{50,b} = \frac{0.87}{0.86}5.4 = 5.46 \text{ in}$$

Since 5.46 inches is greater than the Class 1 D_{50} selected the side slope is also unstable. However, like for the channel bottom, Class 2 (D_{50}= 6 in) would provide a stable side slope.

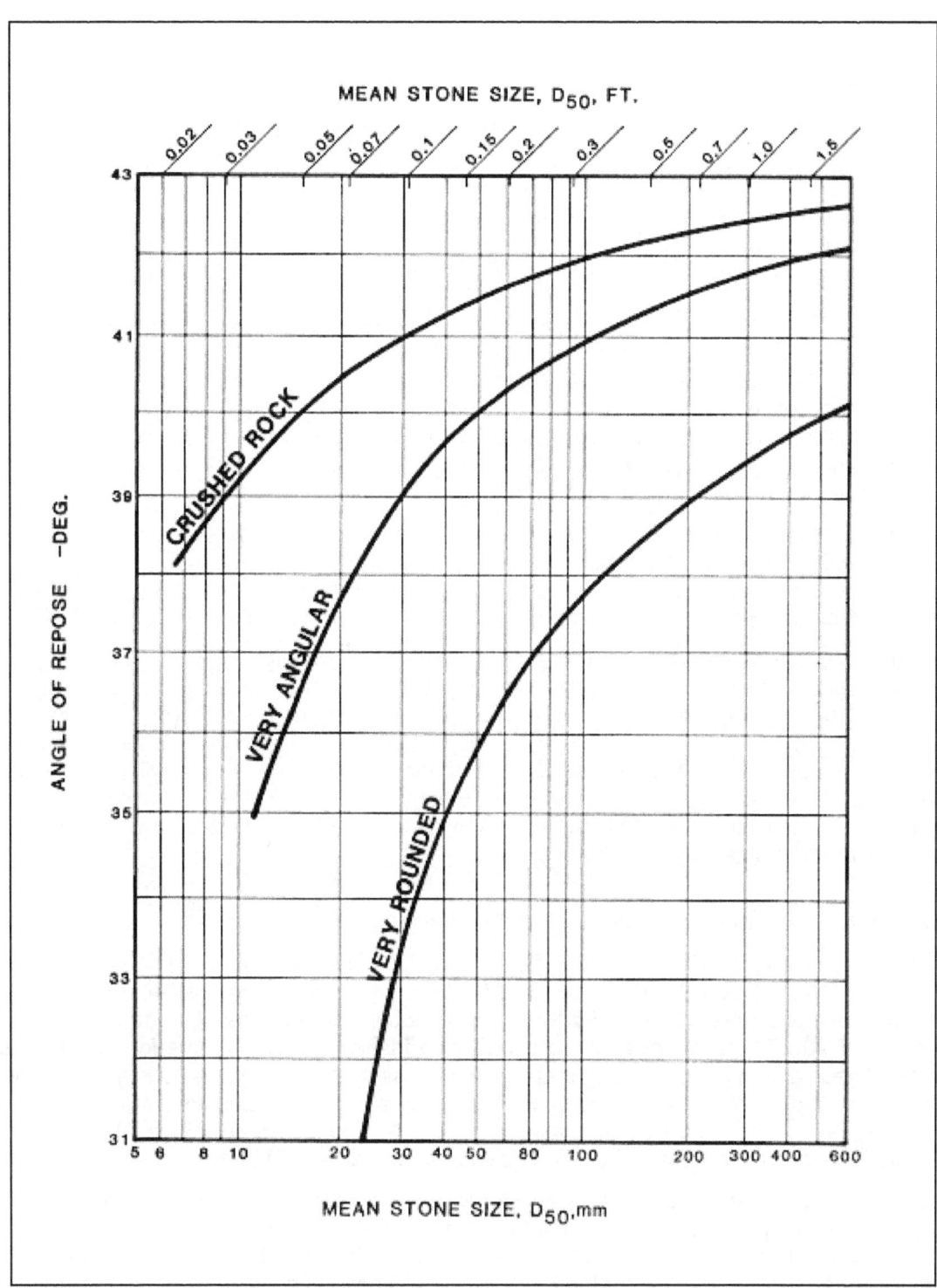

Figure 6.1. Angle of Repose of Riprap in Terms of Mean Size and Shape of Stone

6.3.3 Bends

Added stresses in bends for riprap and gravel linings are treated as described in section 3.4. No additional considerations are required.

6.4 ADDITIONAL CONSIDERATIONS

As with all lining types, the ability to deliver the expected channel protection depends on the proper installation of the lining. Additional design considerations for riprap linings include freeboard; proper specification of gradation and thickness; and use of a filter material under the riprap.

6.4.1 Freeboard and Slope Changes

Freeboard, as defined in Section 2.3.4, is determined based on the predicted water surface elevation in a channel. For channels on mild slopes the water surface elevation for freeboard considerations may be safely taken as the normal depth elevation. For steep slopes and for slope changes, additional consideration of freeboard is required.

For steep channels, freeboard should equal the mean depth of flow, since wave height will reach approximately twice the mean depth. This freeboard height should be used for both transitional and permanent channel installations. Extent of riprap on a steep gradient channel must be sufficient to protect transitions from mild to steep and from steep to mild sections.

The transition from a steep gradient to a culvert should allow for slight movement of riprap. The top of the riprap layer should be placed flush with the invert of a culvert while the riprap layer thickness should equal three to five times the mean rock diameter at the break between the steep slope and culvert entrance. The transition from a steep gradient channel to a mild gradient channel may require an energy dissipation structure such as a plunge pool. The transition from a mild gradient to a steep gradient should be protected against local scour upstream of the transition for a distance of approximately five times the uniform depth of flow in the downstream channel (Chow, 1959).

6.4.2 Riprap Gradation, Angularity, and Thickness

Riprap gradation should follow a smooth size distribution curve. Most riprap gradations will fall in the range of D_{100}/D_{50} and D_{50}/D_{20} between 3.0 to 1.5, which is acceptable. The most important criterion is a proper distribution of sizes in the gradation so that interstices formed by larger stones are filled with smaller sizes in an interlocking fashion, preventing the formation of open pockets. These gradation requirements apply regardless of the type of filter design used. More uniformly graded stone may distribute a higher failure threshold because it contains fewer smaller stones, but at the same time will likely exhibit a more sudden failure. Increasing the safety factor is appropriate when there are questions regarding gradation.

In general, riprap constructed with angular stones has the best performance. Round stones are acceptable as riprap provided they are not placed on side slopes steeper than 1:3 (V:H). Flat slab-like stones should be avoided since they are easily dislodged by the flow. An approximate guide to stone shape is that neither the breadth nor thickness of a single stone is less than one-third its length. Again, the safety factor should be increased if round stones are used. Permissible shear stress estimates are largely based on testing with angular rock.

The thickness of a riprap lining should equal the diameter of the largest rock size in the gradation. For most gradations, this will mean a thickness from 1.5 to 3.0 times the mean riprap diameter. It is important to note that riprap thickness is measured normal to ground surface slope.

6.4.3 Riprap Filter Design

When rock riprap is used, the need for an underlying filter material must be evaluated. The filter material may be either a granular filter blanket or a geotextile fabric.

To determine the need for a filter and, if one is required, to select a gradation for the filter blanket, the following criteria must be met (USACE, 1980). The subscripts "upper" and "lower" refer to the riprap and soil, respectively, when evaluating filter need; the subscripts represent the riprap/filter and filter/soil comparisons when selecting a filter blanket gradation.

$$\frac{D_{15\,upper}}{D_{85\,lower}} < 5 \tag{6.18a}$$

$$5 < \frac{D_{15\,upper}}{D_{15\,lower}} < 40 \tag{6.18b}$$

$$\frac{D_{50\,upper}}{D_{50\,lower}} < 40 \tag{6.18c}$$

In the above relationships, "upper" refers to the overlying material and "lower" refers to the underlying material. The relationships must hold between the filter blanket and base material and between the riprap and filter blanket.

The thickness of the granular filter blanket should approximate the maximum size in the filter gradation. The minimum thickness for a filter blanket should not be less than 150 mm (6 in).

In selecting an engineering filter fabric (geotextile), four properties should be considered (FHWA, 1998):

- Soil retention (piping resistance)
- Permeability
- Clogging
- Survivability

FHWA (1998) provides detailed design guidance for selecting geotextiles as a riprap filter material. These guidelines should be applied in situations where problematic soil environments exist, severe environmental conditions are expected, and/or for critical installations. Problematic soils include unstable or highly erodible soils such as non-cohesive silts; gap graded soils; alternating sand/silt laminated soils; dispersive clays; and/or rock flour. Severe environmental conditions include wave action or high velocity conditions. An installation would be considered critical where loss of life or significant structural damage could be associated with failure.

With the exception of problematic soils or high velocity conditions associated with steep channels and rundowns, geotextile filters for roadside applications may usually be selected based on the apparent opening size (AOS) of the geotextile and the soil type as shown in Table 6.2.

Table 6.2. Maximum AOS for Geotextile Filters (FHWA, 1998)

Soil Type	Maximum AOS (mm)
Non cohesive, less than 15 percent passing the 0.075 mm (US #200) sieve	0.43
Non cohesive, 15 to 50 percent passing the 0.075 mm (US #200) sieve	0.25
Non cohesive, more than 50 percent passing the 0.075 mm (US #200) sieve	0.22
Cohesive, plasticity index greater than 7	0.30

Design Example: Filter Blanket Design

Determine if a granular filter blanket is required, and if so, find an appropriate gradation. Given:

Riprap Gradation

D_{85} = 400 mm (16 in)

D_{50} = 200 mm (8 in)

D_{15} = 100 mm (4in)

Base Soil Gradation

D_{85} =1.5 mm (0.1 in)

D_{50} = 0.5 mm (0.034 in)

D_{15} = 0.167 mm (0.0066 in)

Only an SI solution is provided.

Solution

Check to see if the requirements of Equations 6.18 a, b, and c are met when comparing the riprap (upper) to the underlying soil (lower):

$D_{15\ riprap}/D_{85\ soil} < 5$	substituting	100/1.5 = 67 which is not less than 5
$D_{15\ riprap}/D_{15\ soil} > 5$	substituting	100/0.167 = 600 which is greater than 5, OK
$D_{15\ riprap}/D_{15\ soil} < 40$	substituting	100/0.167 = 600 which is not less than 40
$D_{50\ riprap}/D_{50\ soil} < 40$	substituting	200/0.5 = 400 which is not less than 40

Since three out of the four relationships between riprap and the soil do not meet the recommended dimensional criteria, a filter blanket is required. First, determine the required dimensions of the filter with respect to the base material.

$D_{15\ filter}/D_{85\ soil} < 5$	therefore,	$D_{15\ filter} < 5 \times 1.5$ mm = 7.5 mm
$D_{15\ filter}/D_{15\ soil} > 5$	therefore,	$D_{15\ filter} > 5 \times 0.167$ mm = 0.84 mm
$D_{15\ filter}/D_{15\ soil} < 40$	therefore,	$D_{15\ filter} < 40 \times 0.167$ mm = 6.7 mm
$D_{50\ filter}/D_{50\ soil} < 40$	therefore,	$D_{50\ filter} < 40 \times 0.5$ mm = 20 mm

Therefore, with respect to the soil, the filter must satisfy:

0.84 mm $< D_{15 \text{ filter}} < 6.7$ mm

$D_{50 \text{ filter}} < 20$ mm

Determine the required filter dimensions with respect to the riprap,

$D_{15 \text{ riprap}} / D_{85 \text{ filter}} < 5$	therefore	$D_{85 \text{ filter}} > 100$ mm/5 = 20 mm
$D_{15 \text{ riprap}} / D_{15 \text{ filter}} > 5$	therefore	$D_{15 \text{ filter}} < 100$ mm/5 = 20 mm
$D_{15 \text{ riprap}} / D_{15 \text{ filter}} < 40$	therefore,	$D_{15 \text{ filter}} > 100$ mm/40 = 2.5 mm
$D_{50 \text{ riprap}} / D_{50 \text{ filter}} < 40$	therefore	$D_{50 \text{ filter}} > 200$ mm/40 = 5 mm

With respect to the riprap, the filter must satisfy:

2.5 mm $< D_{15 \text{ filter}} < 20$ mm

$D_{50 \text{ filter}} > 5$ mm

$D_{85 \text{ filter}} > 20$ mm

Combining:

2.5 mm $< D_{15 \text{ filter}} < 6.7$ mm

5 mm $< D_{50 \text{ filter}} < 20$ mm

$D_{85 \text{ filter}} > 20$ mm

A gradation satisfying these requirements is appropriate for this design and is illustrated in Figure 6.2.

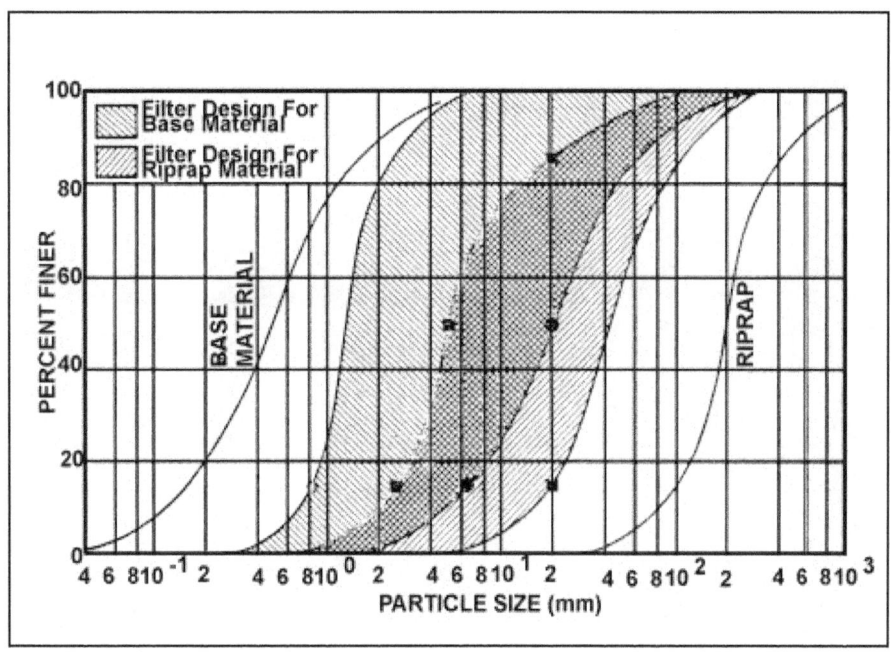

Figure 6.2. Gradations of Granular Filter Blanket for Design Example

This page intentionally left blank.

CHAPTER 7: GABION LINING DESIGN

Gabions (rock filled wire containers) represent an approach for using smaller rock size than would be required by riprap. The smaller rock is enclosed in larger wire units in the form of mattresses or baskets. Gabion baskets are individual rectangular wire mesh containers filled with rock and frequently applied for grade control structures and retaining walls. Gabion mattresses are also rock filled wire mesh containers. The mattresses are composed of a series of integrated cells that hold the rock allowing for a greater spatial extent in each unit. Potential roadside applications for the gabion mattress include steep channels and rundowns.

The thickness of the gabion mattress may be less than the thickness of an equivalently stable riprap lining. Therefore, gabion mattresses represent a trade-off between less and smaller rock versus the costs of providing and installing the wire enclosures. Gabion mattresses are rarely cost effective on mildly sloped channels.

7.1 MANNING'S ROUGHNESS

Roughness characteristics of gabion mattresses are governed by the size of the rock in the baskets and the wire mesh enclosing the rock. For practical purposes, the effect of the mesh can be neglected. Therefore, Manning's roughness should be determined using the D_{50} of the basket rock as applied to the relationships provided for riprap and gravel linings. (See Section 6.1.)

7.2 PERMISSIBLE SHEAR STRESS

Values for permissible shear stress for gabion mattresses are based on research conducted at laboratory facilities and in the field. However, reports from these studies are difficult to reconcile. Simons, et al. (1984) reported permissible shear stresses in the range of 140 to 190 N/m^2 (3 to 4 lb/ft^2) while Clopper and Chen (1988) reported values approaching 1700 N/m^2 (35 lb/ft^2). Simons, et al. tested mattresses ranging in depth from 152 to 457 mm (6 to 18 in) and on slopes of up to 2 percent. Since the objective was to test embankment overtopping, Clopper and Chen tested 152 mm (6 in) mattresses on 25 and 33 percent slopes.

The difference in reported permissible shear stresses may be partly due to the definition of failure. In the Clopper and Chen report, failure was noted after rocks within the basket had shifted to the downstream end of the baskets and an undulating surface was formed leaving part of the embankment exposed. Although this may be an appropriate definition for a rare embankment-overtopping event, such failure is not appropriate for the more frequently occurring roadside design event. For this reason as well as to provide for conservative guidance, the Simons et al. results are emphasized in this guidance.

Permissible shear stress for gabions may be estimated based on the size of the rock fill or based on gabion mattress thickness. Both estimates are determined and the largest value is taken as the permissible shear stress.

Equation 7.1 provides a relationship for permissible shear stress based on rock fill size (Simons, et al., 1984). This shear stress exceeds that of loose riprap because of the added stability provided by the wire mesh. The equation is valid for a range of D_{50} from 0.076 to 0.457 m (0.25 to 1.5 ft)

$$\tau_p = F_*(\gamma_s - \gamma)D_{50}$$

(7.1)

where,

τ_p = permissible shear stress, N/m^2 (lb/ft^2)

F_* = Shields' parameter, dimensionless

D_{50} = median stone size, m (ft)

In the tests reported by Simons, et al. (1984), the Shields' parameter for use in Equation 7.1 was found to be equal to 0.10.

A second equation provides for permissible shear stress based on mattress thickness (Simons, et al., 1984). It is applicable for a range of mattress thickness from 0.152 to 0.457 m (0.5 to 1.5 ft).

$$\tau_p = 0.0091(\gamma_s - \gamma)(MT + MT_c)$$ (7.2)

where,

MT = gabion mattress thickness, m (ft)

MT_C = thickness constant, 1.24 m (4.07 ft)

The limits on Equations 7.1 and 7.2 are based on the range of laboratory data from which they are derived. Rock sizes within mattresses typically range from 0.076 to 0.152 m (0.25 to 0.5 ft) rock in the 0.152 m (0.5 ft) thick mattresses to 0.116 to 0.305 m (0.33 to 1 ft) rock in the 0.457 m (1.5 ft) thick mattresses.

When comparing, the permissible shear for gabions with the calculated shear on the channel, a safety factor, SF is required for Equation 3.2. The guidance found in Table 6.1 is applicable to gabions. Since, the Shields parameter in Equation 7.1 is 0.10, the appropriate corresponding safety factor is 1.25. Alternatively, the designer may compute the particle Reynolds number and, using Table 6.1, determine both a Shields' parameter and SF corresponding to the Reynolds number.

7.3 DESIGN PROCEDURE

The design procedure for gabions is as follows. It uses the same roughness relationships developed for riprap.

Step 1. Determine channel slope, channel shape, and design discharge.

Step 2. Select a trial (initial) mattress thickness and fill rock D_{50}, perhaps based on available sizes for the project. (Also, determine specific weight of proposed stone.)

Step 3. Estimate the depth. For the first iteration, select a channel depth, d_i. For subsequent iterations, a new depth can be estimated from the following equation or any other appropriate method.

$$d_{i+1} = d_i \left(\frac{Q}{Q_i}\right)^{0.4}$$

Determine the average flow depth, d_a in the channel. $d_a = A/T$

Step 4. Calculate the relative depth ratio, d_a/D_{50}. If d_a/D_{50} is greater than or equal to 1.5, use Equation 6.1 to calculate Manning's n. If d_a/D_{50} is less than 1.5 use Equation 6.2 to calculate Manning's n. Calculate the discharge using Manning's equation.

Step 5. If the calculated discharge is within 5 percent of the design discharge, then proceed to step 6. If not, go back to step 3.

Step 6. Calculate the permissible shear stress from Equations 7.1 and 7.2 and take the largest as the permissible shear stress.

Use Equation 3.1 to determine the actual shear stress on the bottom of the channel.

Select a safety factor.

Apply Equation 3.2 to compare the actual to permissible shear stress.

Step 7. If permissible shear is greater than computed shear, the lining is stable. If not, repeat the design process beginning at step 2.

Design Example: Gabion Design (SI)

Determine the flow depth and required thickness of a gabion mattress lining for a trapezoidal channel.

Given:

Q = 0.28 m³/s

S = 0.09 m/m

B = 0.60 m

Z = 3

Solution

Step 1. Channel characteristics and design discharge are given above.

Step 2. Try a 0.23 m thick gabion basket with a D_{50} = 0.15 m; γ_s = 25.9 kN/m³

Step 3. Assume an initial trial depth of 0.3 m

Using the geometric properties of a trapezoid:

$A = Bd + Zd^2 = 0.6(0.3) + 3(0.3)^2 = 0.450 \text{ m}^2$

$P = B + 2d\sqrt{Z^2 + 1} = 0.6 + 2(0.3)\sqrt{3^2 + 1} = 2.50 \text{ m}$

$R = A/P = 0.45/2.50 = 0.180 \text{ m}$

$T = B + 2dZ = 0.6 + 2(0.3)(3) = 2.40 \text{ m}$

$d_a = A/T = 0.45/2.40 = 0.188 \text{ m}$

Step 4. The relative depth ratio, d_a/D_{50} = 0.188/0.150 = 1.3. Therefore, use Equation 6.2 to calculate Manning's n.

$$b = 1.14\left(\frac{D_{50}}{T}\right)^{0.453}\left(\frac{d_a}{D_{50}}\right)^{0.814} = 1.14\left(\frac{0.150}{2.40}\right)^{0.453}\left(\frac{0.188}{0.150}\right)^{0.814} = 0.390$$

$$Fr = \frac{Q/A}{\sqrt{gd_a}} = \frac{0.28/0.450}{\sqrt{9.81(0.188)}} = 0.458$$

$$f(Fr) = \left(\frac{0.28 Fr}{b}\right)^{\log(0.755/b)} = \left(\frac{0.28(0.458)}{0.390}\right)^{\log(0.755/0.389)} = 0.726$$

$$f(REG) = 13.434\left(\frac{T}{D_{50}}\right)^{0.492} b^{1.025(T/D_{50})^{0.118}} = 13.434\left(\frac{2.40}{0.150}\right)^{0.492} 0.390^{1.025(2.40/0.150)^{0.118}} = 13.7$$

$$f(CG) = \left(\frac{T}{d_a}\right)^{-b} = \left(\frac{2.4}{0.188}\right)^{-0.389} = 0.371$$

$$n = \frac{\alpha\, d_a^{1/6}}{\sqrt{g}\ f(Fr)\ f(REG)\ f(CG)} = \frac{1.0\,(0.188)^{1/6}}{\sqrt{9.81}\,(0.726)\,(13.7)(0.371)} = 0.065$$

Calculate Q using Manning's equation:

$$Q = \frac{\alpha}{n} A R^{2/3} S^{1/2} = \frac{1.0}{0.065}(0.45)(0.180)^{2/3}(0.09)^{1/2} = 0.66\ \text{m}^3/\text{s}$$

Step 5. Since this estimate is more than 5 percent from the design discharge, estimate a new depth in step 3.

Step 3 (2nd iteration).   Estimate a new depth estimate:

$$d_{i+1} = d_i \left(\frac{Q}{Q_i}\right)^{0.4} = 0.3\left(\frac{0.28}{0.66}\right)^{0.4} = 0.21\,\text{m}$$

Using the geometric properties of a trapezoid, the maximum and average flow depths are found:

$A = Bd + Zd^2 = 0.6(0.21) + 3(0.21)^2 = 0.258\ \text{m}^2$

$P = B + 2d\sqrt{Z^2 + 1} = 0.6 + 2(0.21)\sqrt{3^2 + 1} = 1.93\ \text{m}$

$R = A/P = 0.258/1.93 = 0.134\ \text{m}$

$T = B + 2dZ = 0.6 + 2(0.21)(3) = 1.86\ \text{m}$

$d_a = A/T = 0.258/1.86 = 0.139\ \text{m}$

Step 4. (2nd iteration). The relative depth ratio, $d_a/D_{50} = 0.139/0.150 = 0.9$. Therefore, use Equation 6.2 to calculate Manning's n.

$$b = 1.14\left(\frac{D_{50}}{T}\right)^{0.453}\left(\frac{d_a}{D_{50}}\right)^{0.814} = 1.14\left(\frac{0.150}{1.86}\right)^{0.453}\left(\frac{0.139}{0.150}\right)^{0.814} = 0.343$$

$$Fr = \frac{Q/A}{\sqrt{gd_a}} = \frac{0.28/0.258}{\sqrt{9.81(0.139)}} = 0.929$$

$$f(Fr) = \left(\frac{0.28Fr}{b}\right)^{log(0.755/b)} = \left(\frac{0.28(0.929)}{0.343}\right)^{log(0.755/0.343)} = 0.910$$

$$f(REG) = 13.434\left(\frac{T}{D_{50}}\right)^{0.492} b^{1.025(T/D_{50})^{0.118}} = 13.434\left(\frac{1.86}{0.150}\right)^{0.492} 0.343^{1.025(1.86/0.150)^{0.118}} = 10.6$$

$$f(CG) = \left(\frac{T}{d_a}\right)^{-b} = \left(\frac{1.86}{0.139}\right)^{-0.343} = 0.411$$

$$n = \frac{\alpha\, d_a^{1/6}}{\sqrt{g}\ f(Fr)\ f(REG)\ f(CG)} = \frac{1.0\,(0.139)^{1/6}}{\sqrt{9.81}\,(0.910)\,(10.6)(0.411)} = 0.058$$

Calculate Q using Manning's equation:

$$Q = \frac{\alpha}{n}AR^{2/3}S^{1/2} = \frac{1.0}{0.058}(0.258)(0.134)^{2/3}(0.09)^{1/2} = 0.35\ m^3/s$$

Since this estimate is also not within 5 percent of the design discharge, further iterations are required. Subsequent iterations will produce the following values:

d = 0.185 m

n = 0.055

Q = 0.29 m³/s

Proceed to step 6 with these values.

Step 6. Calculate the permissible shear stress from Equations 7.1 and 7.2 and take the largest as the permissible shear stress.

$$\tau_p = F_*(\gamma_s - \gamma)D_{50} = 0.10(25,900 - 9810)0.15 = 241\,N/m^2$$

$$\tau_p = 0.0091(\gamma_s - \gamma)(MT + MT_c) = 0.0091(25,900 - 9810)(0.23 + 1.24) = 215\,N/m^2$$

Permissible shear stress for this gabion configuration is, therefore 241 N/m².

Use Equation 3.1 to determine the actual shear stress on the bottom of the channel and apply Equation 3.2 to compare the actual to permissible shear stress.

$$\tau_d = \gamma dS_o = 9810(0.185)(0.09) = 163\,N/m^2$$

SF=1.25:

Step 7. From Equation 3.2: 241>1.25(163), therefore, the selected gabion mattress is acceptable.

Design Example: Gabion Design (CU)

Determine the flow depth and required thickness of a gabion mattress lining for a trapezoidal channel.

Given:

Q   = 10 ft³/s

S = 0.09 ft/ft

B = 2.0 ft

Z = 3

Solution

Step 1. Channel characteristics and design discharge are given above.

Step 2. Try a 0.75 ft thick gabion basket with a D_{50} = 0.5 ft; γ_s= 165 lb/ft^3

Step 3. Assume an initial trial depth of 1 ft.

Using the geometric properties of a trapezoid:

$A = Bd + Zd^2 = 2.0(1.0) + 3(1.0)^2 = 5.0 \text{ ft}^2$

$P = B + 2d\sqrt{Z^2 + 1} = 2.0 + 2(1.0)\sqrt{3^2 + 1} = 8.3 \text{ ft}$

$R = A/P = 5.0/8.3 = 0.601 \text{ ft}$

$T = B + 2dZ = 2.0 + 2(1.0)(3) = 8.0 \text{ ft}$

$d_a = A/T = 5.0/8.0 = 0.625 \text{ ft}$

Step 4. The relative depth ratio, d_a/D_{50} = 0.625/0.50 = 1.3. Therefore, use Equation 6.2 to calculate Manning's n.

$$b = 1.14\left(\frac{D_{50}}{T}\right)^{0.453}\left(\frac{d_a}{D_{50}}\right)^{0.814} = 1.14\left(\frac{0.50}{8.0}\right)^{0.453}\left(\frac{0.625}{0.50}\right)^{0.814} = 0.389$$

$$Fr = \frac{Q/A}{\sqrt{gd_a}} = \frac{10.0/5.0}{\sqrt{32.2(0.625)}} = 0.446$$

$$f(Fr) = \left(\frac{0.28Fr}{b}\right)^{\log(0.755/b)} = \left(\frac{0.28(0.446)}{0.389}\right)^{\log(0.755/0.389)} = 0.721$$

$$f(REG) = 13.434\left(\frac{T}{D_{50}}\right)^{0.492} b^{1.025(T/D_{50})^{0.118}} = 13.434\left(\frac{8.0}{0.50}\right)^{0.492} 0.389^{1.025(8.0/0.50)^{0.118}} = 13.7$$

$$f(CG) = \left(\frac{T}{d_a}\right)^{-b} = \left(\frac{8.0}{0.50}\right)^{-0.389} = 0.371$$

$$n = \frac{\alpha\, d_a^{1/6}}{\sqrt{g}\ f(Fr)\ f(REG)\ f(CG)} = \frac{1.49\,(0.625)^{1/6}}{\sqrt{32.2}\ (0.721)\ (13.7)(0.371)} = 0.066$$

Calculate Q using Manning's equation:

$$Q = \frac{\alpha}{n} AR^{2/3}S^{1/2} = \frac{1.49}{0.066}(5.00)(0.601)^{2/3}(0.09)^{1/2} = 24.1 \text{ ft}^3/\text{s}$$

Step 5. Since this estimate is more than 5 percent from the design discharge, estimate a new depth in step 3.

Step 3 (2nd iteration). Estimate a new depth estimate:

$$d_{i+1} = d_i \left(\frac{Q}{Q_i}\right)^{0.4} = 1.0 \left(\frac{10}{19.8}\right)^{0.4} = 0.70 \text{ ft}$$

Using the geometric properties of a trapezoid, the maximum and average flow depths are found:

$A = Bd+Zd^2 = 2.0(0.70)+3(0.70)^2 = 2.87 \text{ ft}^2$

$P = B + 2d\sqrt{Z^2 + 1} = 2.0 + 2(0.70)\sqrt{3^2 + 1} = 6.43 \text{ ft}$

$R = A/P = 2.87/6.43 = 0.446 \text{ ft}$

$T = B+2dZ = 2.0+2(0.70)(3) = 6.20 \text{ ft}$

$d_a = A/T = 2.87/6.20 = 0.463 \text{ ft}$

Step 4. (2nd iteration). The relative depth ratio, $d_a/D_{50} = 0.496/0.50 = 1.0$. Therefore, use Equation 6.2 to calculate Manning's n.

$$b = 1.14 \left(\frac{D_{50}}{T}\right)^{0.453} \left(\frac{d_a}{D_{50}}\right)^{0.814} = 1.14 \left(\frac{0.50}{6.20}\right)^{0.453} \left(\frac{0.463}{0.50}\right)^{0.814} = 0.342$$

$$Fr = \frac{Q/A}{\sqrt{gd_a}} = \frac{10.0/2.87}{\sqrt{32.2(0.463)}} = 0.902$$

$$f(Fr) = \left(\frac{0.28Fr}{b}\right)^{\log(0.755/b)} = \left(\frac{0.28(0.902)}{0.342}\right)^{\log(0.755/0.342)} = 0.901$$

$$f(REG) = 13.434 \left(\frac{T}{D_{50}}\right)^{0.492} b^{1.025(T/D_{50})^{0.118}} = 13.434 \left(\frac{6.20}{0.50}\right)^{0.492} 0.342^{1.025(6.20/0.50)^{0.118}} = 10.6$$

$$f(CG) = \left(\frac{T}{d_a}\right)^{-b} = \left(\frac{6.20}{0.463}\right)^{-0.389} = 0.412$$

$$n = \frac{\alpha \, d_a^{1/6}}{\sqrt{g} \, f(Fr) \, f(REG) \, f(CG)} = \frac{1.49 \, (0.463)^{1/6}}{\sqrt{32.2} \, (0.901) \, (10.6)(0.412)} = 0.059$$

Calculate Q using Manning's equation:

$$Q = \frac{\alpha}{n} AR^{2/3}S^{1/2} = \frac{1.49}{0.059}(2.87)(0.446)^{2/3}(0.09)^{1/2} = 12.7 \text{ ft}^3/s$$

Step 5 (2nd iteration). Since this estimate is also not within 5 percent of the design discharge, further iterations are required. Subsequent iterations will produce the following values:

d = 0.609 ft

n = 0.055

Q = 10.2 ft³/s

Proceed to step 6 with these values.

Step 6. Calculate the permissible shear stress from Equations 7.1 and 7.2 and take the largest as the permissible shear stress.

$$\tau_p = F_*(\gamma_s - \gamma)D_{50} = 0.10(165 - 62.4)0.5 = 5.1\,lb/ft^2$$

$$\tau_p = 0.0091(\gamma_s - \gamma)(MT + MT_c) = 0.0091(165 - 62.4)(0.75 + 4.07) = 4.5\,lb/ft^2$$

Permissible shear stress for this gabion configuration is, therefore 5.1 lb/ft².

Use Equation 3.1 to determine the actual shear stress on the bottom of the channel and apply Equation 3.2 to compare the actual to permissible shear stress.

$$\tau_d = \gamma dS_o = 62.4(0.609)(0.09) = 3.4\,lb/ft^2$$

SF=1.25:

Step 7. From Equation 3.2: 5.1>1.25(3.4), therefore, the selected gabion mattress is acceptable.

7.4 ADDITIONAL CONSIDERATIONS

As with riprap linings, the ability to deliver the expected channel protection depends on the proper installation of the lining. Additional design considerations for gabion linings include consideration of the wire mesh; freeboard; proper specification of gradation and thickness; and use of a filter material under the gabions.

The stability of gabions depends on the integrity of the wire mesh. In streams with high sediment concentrations or with rocks moving along the bed of the channel, the wire mesh may be abraded and eventually fail. Under these conditions the gabion will no longer behave as a single unit but rather as individual stones. Applications of gabion mattresses and baskets under these conditions should be avoided. Such conditions are unlikely for roadside channel design.

Extent of gabions on a steep gradient (the most common roadside application for gabions) must be sufficient to protect transition regions of the channel both above and below the steep gradient section. The transition from a steep gradient to a culvert should allow for slumping of a gabion mattress.

Gabions should be placed flush with the invert of a culvert. The break between the steep slope and culvert entrance should equal three to five times the mattress thickness. The transition from a steep gradient channel to a mild gradient channel may require an energy dissipation structure such as a plunge pool. The transition from a mild gradient to a steep gradient should be protected against local scour upstream of the transition for a distance of approximately five times the uniform depth of flow in the downstream channel (Chow, 1959).

Freeboard should equal the mean depth of flow, since wave height will reach approximately twice the mean depth. This freeboard height should be used for both transitional and permanent channel installations.

The rock gradation used in gabions mattress must be such that larger stones do not protrude outside the mattress and the wire mesh retains smaller stones.

When gabions are used, the need for an underlying filter material must be evaluated. The filter material may be either a granular filter blanket or geotextile fabric. See section 6.4.3 for description of the filter requirements.

This page intentionally left blank.

APPENDIX A: METRIC SYSTEM, CONVERSION FACTORS, AND WATER PROPERTIES

The following information is summarized from the Federal Highway Administration, National Highway Institute (NHI) Course No. 12301, "Metric (SI) Training for Highway Agencies." For additional information, refer to the Participant Notebook for NHI Course No. 12301.

In SI there are seven base units, many derived units and two supplemental units (Table A.1). Base units uniquely describe a property requiring measurement. One of the most common units in civil engineering is length, with a base unit of meters in SI. Decimal multiples of meter include the kilometer (1000 m), the centimeter (1m/100) and the millimeter (1 m/1000). The second base unit relevant to highway applications is the kilogram, a measure of mass that is the inertia of an object. There is a subtle difference between mass and weight. In SI, mass is a base unit, while weight is a derived quantity related to mass and the acceleration of gravity, sometimes referred to as the force of gravity. In SI the unit of mass is the kilogram and the unit of weight/force is the newton. Table A.2 illustrates the relationship of mass and weight. The unit of time is the same in SI as in the Customary (English) system (seconds). The measurement of temperature is Centigrade. The following equation converts Fahrenheit temperatures to Centigrade, $°C = 5/9 (°F - 32)$.

Derived units are formed by combining base units to express other characteristics. Common derived units in highway drainage engineering include area, volume, velocity, and density. Some derived units have special names (Table A.3).

Table A.4 provides useful conversion factors from Customary to SI units. The symbols used in this table for metric (SI) units, including the use of upper and lower case (e.g., kilometer is "km" and a newton is "N") are the standards that should be followed. Table A.5 provides the standard SI prefixes and their definitions.

Table A.6 provides physical properties of water at atmospheric pressure in SI units. Table A.7 gives the sediment grade scale and Table A.8 gives some common equivalent hydraulic units.

Table A.1. Overview of SI Units

	Units	Symbol

Base units
    length
    mass
    time
    temperature*
    electrical current
    luminous intensity
    amount of material
	
	meter
	kilogram
	second
	kelvin
	ampere
	candela
	mole
		
		m
		kg
		s
		K
		A
		cd
		mol

Derived units		

Supplementary units
    angles in the plane
    solid angles
	
	radian
	steradian
		
		rad
		sr

*Use degrees Celsius (°C), which has a more common usage than kelvin.		

Table A.2. Relationship of Mass and Weight

	Mass	Weight or Force of Gravity	Force
Customary	slug pound-mass	pound pound-force	pound pound-force
Metric	kilogram	newton	newton

Table A.3. Derived Units With Special Names

Quantity	Name	Symbol	Expression
Frequency	hertz	Hz	s^{-1}
Force	newton	N	$kg \cdot m/s^2$
Pressure, stress	pascal	Pa	N/m^2
Energy, work, quantity of heat	joule	J	N•m
Power, radiant flux	watt	W	J/s
Electric charge, quantity	coulomb	C	A•s
Electric potential	volt	V	W/A
Capacitance	farad	F	C/V
Electric resistance	ohm	Ω	V/A
Electric conductance	siemens	S	A/V
Magnetic flux	weber	Wb	V•s
Magnetic flux density	tesla	T	Wb/m^2
Inductance	henry	H	Wb/A
Luminous flux	lumen	lm	cd•sr
Illuminance	lux	lx	lm/m^2

Table A.4. Useful Conversion Factors

Quantity	From Customary Units	To Metric Units	Multiplied By*
Length	mile	Km	1.609
	yard	m	0.9144
	foot	m	<u>0.3048</u>
	inch	mm	<u>25.40</u>
Area	square mile	km^2	2.590
	acre	m^2	4047
	acre	hectare	0.4047
	square yard	m^2	0.8361
	square foot	m^2	0.09290
	square inch	mm^2	645.2
Volume	acre foot	m^3	1233
	cubic yard	m^3	0.7646
	cubic foot	m^3	0.02832
	cubic foot	L (1000 cm^3)	28.32
	100 board feet	m^3	0.2360
	gallon	L (1000 cm^3)	3.785
	cubic inch	cm^3	16.39
Mass	lb	kg	0.4536
	kip (1000 lb)	metric ton (1000 kg)	0.4536
Mass/unit length	plf	kg/m	1.488
Mass/unit area	psf	kg/m^2	4.882
Mass density	pcf	kg/m^3	16.02
Force	lb	N	4.448
	kip	kN	4.448
Force/unit length	plf	N/m	14.59
	klf	kN/m	14.59
Pressure, stress, modulus of elasticity	psf	Pa	47.88
	ksf	kPa	47.88
	psi	kPa	6.895
	ksi	MPa	6.895
Bending moment, torque, moment of force	ft-lb	N A m	1.356
	ft-kip	kN A m	1.356
Moment of mass	lb•ft	m	0.1383
Moment of inertia	lb•ft^2	kg•m^2	0.04214
Second moment of area	In^4	mm^4	416200
Section modulus	in^3	mm^3	16390
Power	ton (refrig)	kW	3.517
	Btu/s	kW	1.054
	hp (electric)	W	745.7
	Btu/h	W	0.2931
Volume rate of flow	ft^3/s	m^3/s	0.02832
	cfm	m^3/s	0.0004719
	cfm	L/s	0.4719
	mgd	m^3/s	0.0438
Velocity, speed	ft/s	M/s	<u>0.3048</u>
Acceleration	F/s^2	m/s^2	<u>0.3048</u>
Momentum	lb•ft/sec	kg•m/s	0.1383
Angular momentum	lb•ft^2/s	kg•m^2/s	0.04214
Plane angle	degree	rad	0.01745
		mrad	17.45

*4 significant figures; underline denotes exact conversion

Table A.5. Prefixes

Submultiples			Multiples		
Deci	10^{-1}	d	deka	10^{1}	da
Centi	10^{-2}	c	hecto	10^{2}	h
Milli	10^{-3}	m	kilo	10^{3}	k
Micro	10^{-6}	μ	mega	10^{6}	M
Nano	10^{-9}	n	giga	10^{9}	G
Pica	10^{-12}	p	tera	10^{12}	T
Femto	10^{-15}	f	peta	10^{15}	P
Atto	10^{-18}	a	exa	10^{18}	E
Zepto	10^{-21}	z	zetta	10^{21}	Z
Yocto	10^{-24}	y	yotto	10^{24}	Y

Table A.6. Physical Properties of Water at Atmospheric Pressure in SI Units.

Temperature		Density	Specific Weight	Dynamic Viscosity	Kinematic Viscosity	Vapor Pressure	Surface Tension[1]	Bulk Modulus
Centigrade	Fahrenheit	kg/m³	N/m³	N · s/m²	m²/s	N/m² abs.	N/m	GN/m²
0°	32°	1,000	9,810	1.79×10^{-3}	1.79×10^{-6}	611	0.0756	1.99
5°	41°	1,000	9,810	1.51×10^{-3}	1.51×10^{-6}	872	0.0749	2.05
10°	50°	1,000	9,810	1.31×10^{-3}	1.31×10^{-6}	1,230	0.0742	2.11
15°	59°	999	9,800	1.14×10^{-3}	1.14×10^{-6}	1,700	0.0735	2.16
20°	68°	998	9,790	1.00×10^{-3}	1.00×10^{-6}	2,340	0.0728	2.20
25°	77°	997	9,781	8.91×10^{-4}	8.94×10^{-7}	3,170	0.0720	2.23
30°	86°	996	9,771	7.97×10^{-4}	8.00×10^{-7}	4,250	0.0712	2.25
35°	95°	994	9,751	7.20×10^{-4}	7.24×10^{-7}	5,630	0.0704	2.27
40°	104°	992	9,732	6.53×10^{-4}	6.58×10^{-7}	7,380	0.0696	2.28
50°	122°	988	9,693	5.47×10^{-4}	5.53×10^{-7}	12,300	0.0679	
60°	140°	983	9,643	4.66×10^{-4}	4.74×10^{-7}	20,000	0.0662	
70°	158°	978	9,594	4.04×10^{-4}	4.13×10^{-7}	31,200	0.0644	
80°	176°	972	9,535	3.54×10^{-4}	3.64×10^{-7}	47,400	0.0626	
90°	194°	965	9,467	3.15×10^{-4}	3.26×10^{-7}	70,100	0.0607	
100°	212°	958	9,398	2.82×10^{-4}	2.94×10^{-7}	101,300	0.0589	

[1]Surface tension of water in contact with air

Table A.7. Physical Properties of Water at Atmospheric Pressure in English Units.

Temperature Fahrenheit	Temperature Centigrade	Density Slugs/ft^3	Specific Weight Weight lb/ft^3	Dynamic Viscosity lb-sec/ft^2	Kinematic Viscosity ft^2/sec	Vapor Pressure lb/in^2	Surface Tension[1] lb/ft	Bulk Modulus lb/in^2
32	0	1.940	62.416	0.374×10^{-4}	1.93×10^{-5}	0.09	0.00518	287,000
39.2	4.0	1.940	62.424					
40	4.4	1.940	62.423	0.323	1.67	0.12	.00514	296,000
50	10.0	1.940	62.408	0.273	1.41	0.18	.00508	305,000
60	15.6	1.939	62.366	0.235	1.21	0.26	.00504	313,000
70	21.1	1.936	62.300	0.205	1.06	0.36	.00497	319,000
80	26.7	1.934	62.217	0.180	0.929	0.51	.00492	325,000
90	32.2	1.931	62.118	0.160	0.828	0.70	.00486	329,000
100	37.8	1.927	61.998	0.143	0.741	0.95	.00479	331,000
120	48.9	1.918	61.719	0.117	0.610	1.69	.00466	332,000
140	60.0	1.908	61.386	0.0979	0.513	2.89		
160	71.1	1.896	61.006	0.0835	0.440	4.74		
180	82.2	1.883	60.586	0.0726	0.385	7.51		
200	93.3	1.869	60.135	0.0637	0.341	11.52		
212	100	1.847	59.843	0.0593	0.319	14.70		

[1]Surface tension of water in contact with air

A-7

Table A.8. Sediment Particles Grade Scale.

Size				Approximate Sieve Mesh Openings Per Inch		Class
Millimeters		Microns	Inches	Tyler	U.S. Standard	
4000-2000	-----	-----	160-80	-----	-----	Very large boulders
2000-1000	-----	-----	80-40	-----	-----	Large boulders
1000-500	-----	-----	40-20	-----	-----	Medium boulders
500-250	-----	-----	20-10	-----	-----	Small boulders
250-130	-----	-----	10-5	-----	-----	Large cobbles
130-64	-----	-----	5-2.5	-----	-----	Small cobbles
64-32	-----	-----	2.5-1.3	-----	-----	Very coarse gravel
32-16	-----	-----	1.3-0.6	-----	-----	Coarse gravel
16-8	-----	-----	0.6-0.3	2 1/2	-----	Medium gravel
8-4	-----	-----	0.3-0.16	5	5	Fine gravel
4-2	-----	-----	0.16-0.08	9	10	Very fine gravel
2-1	2.00-1.00	2000-1000	-----	16	18	Very coarse sand
1-1/2	1.00-0.50	1000-500	-----	32	35	Coarse sand
1/2-1/4	0.50-0.25	500-250	-----	60	60	Medium sand
1/4-1/8	0.25-0.125	250-125	-----	115	120	Fine sand
1/8-1/16	0.125-0.062	125-62	-----	250	230	Very fine sand
1/16-1/32	0.062-0.031	62-31	-----	-----	-----	Coarse silt
1/32-1/64	0.031-0.016	31-16	-----	-----	-----	Medium silt
1/64-1/128	0.016-0.008	16-8	-----	-----	-----	Fine silt
1/128-1/256	0.008-0.004	8-4	-----	-----	-----	Very fine silt
1/256-1/512	0.004-0.0020	4-2	-----	-----	-----	Coarse clay
1/512-1/1024	0.0020-0.0010	2-1	-----	-----	-----	Medium clay
1/1024-1/2048	0.0010-0.0005	1-0.5	-----	-----	-----	Fine clay
1/2048-1/4096	0.0005-0.0002	0.5-0.24	-----	-----	-----	Very fine clay

Table A.9. Common Equivalent Hydraulic Units.

Volume

Unit	Equivalent							
	cubic inch	liter	u.s. gallon	cubic foot	cubic yard	cubic meter	acre-foot	sec-foot-day
liter	61.02	1	0.264 2	0.035 31	0.001 308	0.001	810.6 E - 9	408.7 E - 9
U.S. gallon	231.0	3.785	1	0.133 7	0.004 951	0.003 785	3.068 E - 6	1.547 E - 6
cubic foot	1728	28.32	7.481	1	0.037 04	0.028 32	22.96 E - 6	11.57 E - 6
cubic yard	46 660	764.6	202.0	27	1	0.746 6	619.8 E - 6	312.5 E - 6
$meter^3$	61 020	1000	264.2	35.31	1.308	1	810.6 E - 6	408.7 E - 6
acre-foot	75.27 E + 6	1 233 000	325 900	43 560	1 613	1 233	1	0.504 2
sec-foot-day	149.3 E + 6	2 447 000	646 400	86 400	3 200	2 447	1.983	1

Discharge (Flow Rate, Volume/Time)

Unit	Equivalent					
	gallon/min	liter/sec	acre-foot/day	$foot^3$/sec	million gal/day	$meter^3$/sec
gallon/minute	1	0.063 09	0.004 419	0.002 228	0.001 440	63.09 E - 6
liter/second	15.85	1	0.070 05	0.035 31	0.022 82	0.001
acre-foot/day	226.3	14.28	1	0.504 2	0.325 9	0.014 28
$feet^3$/second	448.8	28.32	1.983	1	0.646 3	0.028 32
million gal/day	694.4	43.81	3.068	1.547	1	0.043 82
$meter^3$/second	15 850	1000	70.04	35.31	22.82	1

This page intentionally left blank.

APPENDIX B: CHANNEL GEOMETRY EQUATIONS

V- SHAPE

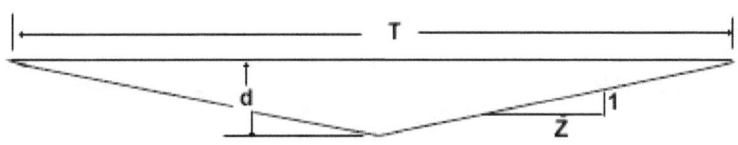

$$A = Zd^2$$
$$p = 2d\sqrt{z^2 + 1}$$
$$T = 2dZ$$

TRAPEZOIDAL

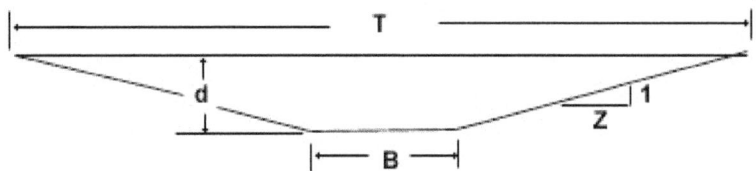

$$A = Bd + Zd^2$$
$$P = B + 2d\sqrt{z^2 + 1}$$
$$T = B + 2dZ$$

PARABOLIC

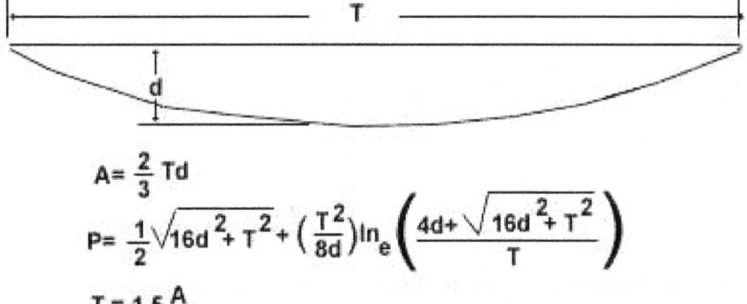

$$A = \frac{2}{3}Td$$
$$P = \frac{1}{2}\sqrt{16d^2 + T^2} + \left(\frac{T^2}{8d}\right)\ln_e\left(\frac{4d + \sqrt{16d^2 + T^2}}{T}\right)$$
$$T = 1.5\frac{A}{d}$$

V-SHAPE WITH ROUNDED BOTTOM

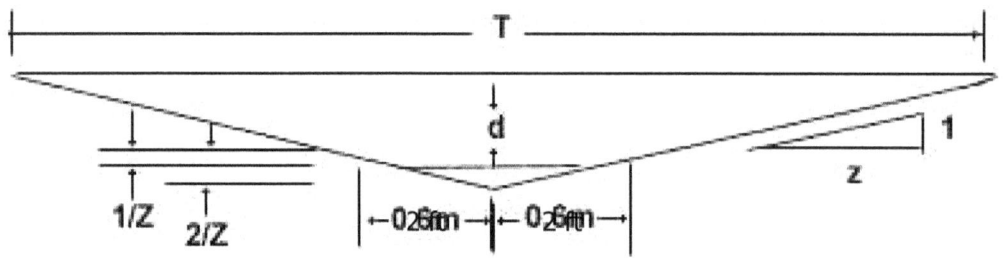

2 CASES

No. 1

If d ≤ 1/Z, then:

$$A = \frac{8}{3}d\sqrt{dZ}$$

$$P = 2Z\ln_e\left(\sqrt{\frac{d}{Z}} + \sqrt{1+\frac{d}{Z}}\right) + 2\sqrt{d^2 + dZ}$$

$$T = 4\sqrt{dZ}$$

No. 2

If d > 1/Z, then:

$$A = \frac{8}{3}d + 4\left(d - \frac{1}{Z}\right) + Z\left(d - \frac{1}{Z}\right)^2$$

$$P = 2Z\ln_e\left(\frac{1+\sqrt{Z^2+1}}{Z}\right) + 2\frac{\sqrt{Z^2+1}}{Z} + 2\left(d-\frac{1}{Z}\right)\sqrt{1+Z^2}$$

$$T = 4 + 2Z\left(d - \frac{1}{Z}\right)$$

Note: The equations for V-shape with rounded bottom only apply in customary units for a channel with a 4 ft wide rounded bottom.

B -2

APPENDIX C: RESISTANCE EQUATIONS

C.1 GENERAL RELATIONSHIPS

Resistance to flow in open channels with flexible linings can be accurately described using the universal-velocity-distribution law (Chow, 1959). The form of the resulting equation is:

$$V = V_* \left[a + b \ \log\left(\frac{R}{k_s}\right) \right] \quad\quad\quad\text{(C.1)}$$

where,

V = mean channel velocity, m/s (ft/s)

V_* = shear velocity which is $\sqrt{gRS_f}$, m/s (ft/s)

a, b = empirical coefficients

R = hydraulic radius, m (ft)

k_S = roughness element height, m (ft)

g = acceleration due to gravity, m/s^2 (ft/s^2)

Manning's equation and Equation C.1 can be combined to give Manning's roughness coefficient n in terms of the relative roughness. The resulting equation is:

$$n = \frac{\alpha \ R^{1/6}}{\sqrt{g}\left(a + b \ \log\left(\dfrac{R}{k_s}\right) \right)} \quad\quad\quad\text{(C.2)}$$

where,

α = unit conversion constant, 1.0 (SI) and 1.49 (CU)

C.2   GRASS LINING FLOW RESISTANCE

General form of the relative roughness equation (Kouwen and Unny, 1969; Kouwen and Li, 1981) is as follows.

$$\sqrt{\frac{1}{f}} = a + b \cdot \log\left(\frac{d}{k}\right) \qquad\qquad (C.3)$$

where,

f	=	Darcy-Weisbach friction factor
a, b	=	parameters for relative roughness formula
d	=	depth of flow, m (ft)
k	=	roughness height (deflected height of grass stem), m (ft)

Coefficients "a" and "b" are a function of shear velocity, V_*, relative to critical shear velocity, V_{*crit}, where the critical shear velocity is a function of the density-stiffness property of the grass stem. Table C.1 provides upper and lower limits of the relative roughness coefficients. Within these limits, values of the coefficients can be estimated by linear interpolation on V_*/V_{*crit}.

Table C.1. Resistance Equation Coefficients

V_*/V_{*crit}	a	b
1.0	0.15	1.85
2.5	0.29	3.50

Critical shear velocity is estimated as the minimum value computed from the following two equations:

$$V_{*crit} = \alpha_1 + \alpha_2 MEI^2 \qquad\qquad (C.4a)$$

$$V_{*crit} = \alpha_3 MEI^{0.106} \qquad\qquad (C.4b)$$

where,

V_{*crit}	=	critical shear velocity, m/s (ft/s)
MEI	=	density-stiffness parameter, N•m² (lb•ft²)
α_1	=	unit conversion constant, 0.028 (SI) and 0.092 (CU)
α_2	=	unit conversion constant, 6.33 (SI) and 3.55 (CU)
α_3	=	unit conversion constant, 0.23 (SI) and 0.69 (CU)

Note: For MEI < 0.16 N•m² (0.39 lb•ft²) the second equation controls, which is in the D to E retardance range.

The roughness height, k, is a function of density, M, and stiffness (EI) parameter (MEI). Stiffness is the product of modulus of elasticity (E) and second moment of stem cross sectional area (I).

$$\frac{k}{h} = 0.14\left(\frac{(MEI/\tau_o)^{1/4}}{h}\right)^{8/5}$$

(C.5)

where,

h = grass stem height, m (ft)

τ_o = mean boundary shear stress, N/m^2, lb/ft^2

Values of h and MEI for various classifications of vegetative roughness, known as retardance classifications, are given in Table C.2.

Table C.2. Relative Roughness Parameters for Vegetation

Retardance Class	Average Height, h		Density-stiffness, MEI	
	m	Ft	N•m^2	lb•ft^2
A	0.91	3.0	300	725
B	0.61	2.0	20	50
C	0.20	0.66	0.5	1.2
D	0.10	0.33	0.05	0.12
E	0.04	0.13	0.005	0.012

Eastgate (1966) showed a relationship between a fall-board test (Appendix E) and the MEI property as follows:

$$MEI = \alpha h_b^{2.82}$$

(C.6)

where,

h_b = deflected grass stem height resulting from the fall-board test, m (ft)

α = unit conversion constant, 3120 (SI) and 265 (CU)

Kouwen collected additional data using the fall-board test (Kouwen, 1988). These data have been interpreted to have the following relationship.

$$MEI = C_s h^{2.82}$$

(C.7)

where,

C_s = grass density-stiffness coefficient

Combining Equations C.5 and C.7 gives:

$$k = 0.14 C_s^{0.4} h^{0.528} \left(\frac{1}{\tau_o}\right)^{0.4}$$

(C.8)

Over a range of shallow depths ($y \leq 0.9$ m (3 ft)), the n value is a function of roughness height as shown in Figure C.1. The linear relationships shown between roughness height, k, and Manning's n differ with vegetation condition.

$$n = \alpha C_l k \qquad (C.9)$$

where,

C_l = k-n coefficient

α = unit conversion constant, 1.0 (SI) and 0.213 (CU)

As is apparent in Figure C.2, a relationship exists between C_l and C_s and is quantified in the following equation:

$$C_l = 2.5C_s^{-0.3} \qquad (C.10)$$

Substituting Equations C.10 and C.8 in Equation C.9 yields the following:

$$n = \alpha(0.35)C_s^{0.1}h^{0.528}\left(\frac{1}{\tau_o}\right)^{0.4} \qquad (C.11)$$

Defining a grass roughness coefficient, C_n as,

$$C_n = 0.35C_s^{0.10}h^{0.528} \qquad (C.12)$$

yields the following relationship for Manning's n:

$$n = \alpha \frac{C_n}{\tau_o^{0.4}} \qquad (C.13)$$

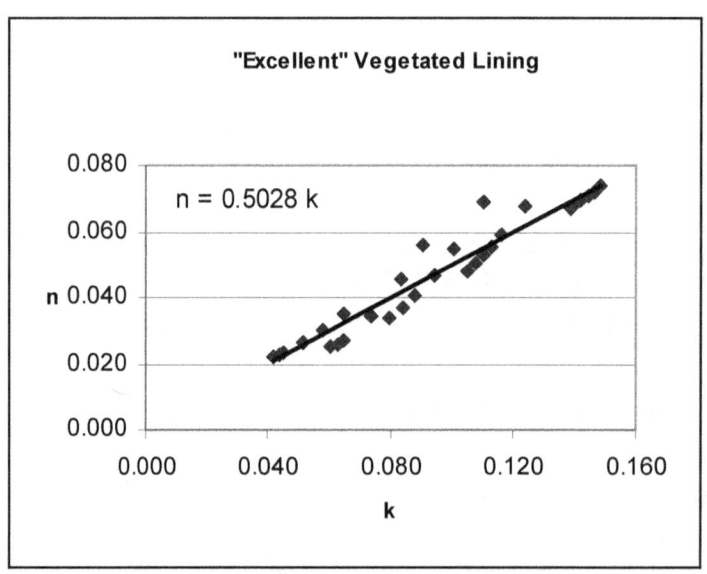

Figure C.1a. Relative Roughness Relationships for Excellent Vegetated Conditions

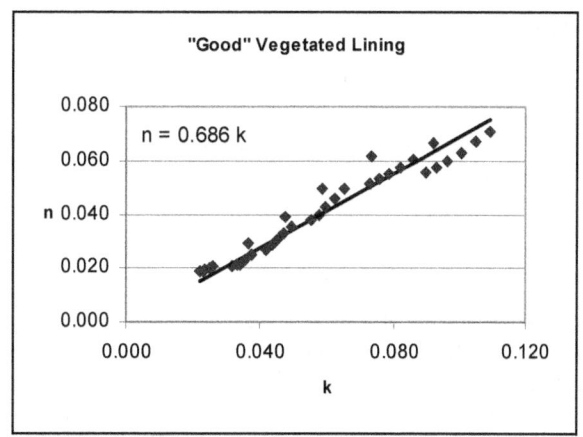

Figure C.1b. Relative Roughness Relationships for Good Vegetated Conditions

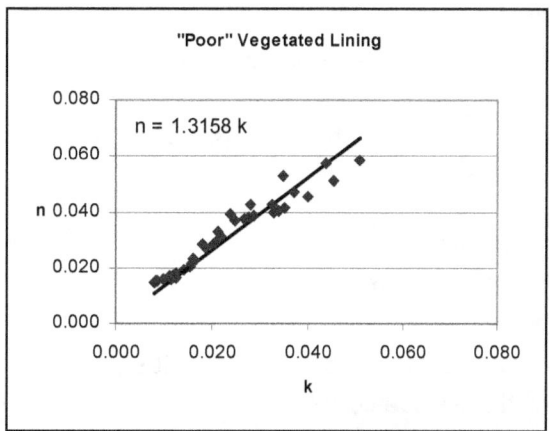

Figure C.1c. Relative Roughness Relationships for Poor Vegetated Conditions

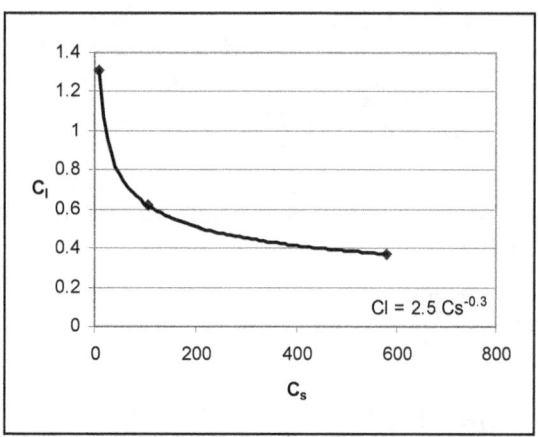

Figure C.2. Relationship between C_l and C_s

C -5

C.3 BATHURST RESISTANCE EQUATION

Most of the flow resistance in channels with large-scale (relative to depth) roughness is derived from the form drag of the roughness elements and the distortion of the flow as it passes around roughness elements. Consequently, a flow resistance equation for these conditions has to account for skin friction and form drag. Because of the shallow depths of flow and the large size of the roughness elements, the flow resistance will vary with relative roughness area, roughness geometry, Froude number (the ratio of inertial forces to gravitational forces), and Reynolds number (the ratio of inertial forces to viscous forces).

Bathurst's experimental work quantified these relationships in a semi-empirical fashion. The work shows that for Reynolds numbers in the range of 4×10^4 to 2×10^5, resistance is likely to fall significantly as Reynolds number increases. For Reynolds numbers in excess of 2×10^5, the Reynolds effect on resistance remains constant. When roughness elements protrude through the free surface, resistance increases significantly due to Froude number effects, i.e., standing waves, hydraulic jumps, and free-surface drag. For the channel as a whole, free-surface drag decreases as the Froude number and relative submergence increase. Once the elements are submerged, Froude number effects related to free-surface drag are small, but those related to standing waves are important.

The general dimensionless form of the Bathurst equation is:

$$\frac{V}{V_*} = \frac{\alpha d_a^{1/6}}{n\sqrt{g}} = f(Fr)\, f(REG)\, f(CG) \tag{C.14}$$

where,

V	=	mean velocity, m/s (ft/s)
V_*	=	shear velocity = $(gdS)^{0.5}$, m/s (ft/s)
d_a	=	mean flow depth, m (ft)
g	=	acceleration due to gravity, 9.81 m/s^2 (32.2 ft/s^2)
n	=	Manning's roughness coefficient
Fr	=	Froude number
REG	=	roughness element geometry
CG	=	channel geometry
α	=	unit conversion constant, 1.0 (SI) and 1.49 (CU)

Equation C.14 can be rewritten in the following form to describe the relationship for n.

$$n = \frac{\alpha\, d_a^{1/6}}{\sqrt{g}\, f(Fr)\, f(REG)\, f(CG)} \tag{C.15}$$

The functions of Froude number, roughness element geometry, and channel geometry are given by the following equations:

$$f(Fr) = \left(\frac{0.28Fr}{b}\right)^{\log(0.755/b)} \tag{C.16}$$

$$f(REG) = 13.434 \left(\frac{T}{Y_{50}}\right)^{0.492} b^{1.025(T/Y_{50})^{0.118}} \tag{C.17}$$

C -6

$$f(CG) = \left(\frac{T}{d_a}\right)^{-b}$$ (C.18)

where,

T = channel top width, m (ft)

Y_{50} = mean value of the distribution of the average of the long and median axes of a roughness element, m (ft)

b = parameter describing the effective roughness concentration.

The parameter b describes the relationship between effective roughness concentration and relative submergence of the roughness bed. This relationship is given by:

$$b = a\left(\frac{d_a}{S_{50}}\right)^c$$ (C.19)

where,

S_{50} = mean of the short axis lengths of the distribution of roughness elements, m (ft)

a, c = constants varying with bed material properties.

The parameter, c, is a function of the roughness size distribution and varies with respect to the bed-material gradation. σ, where:

$$c = 0.648\sigma^{-0.134}$$ (C.20)

For standard riprap gradations the log standard deviation is assumed to be constant at a value of 0.182, giving a c value of 0.814.

The parameter, a, is a function of channel width and bed material size in the cross stream direction, and is defined as:

$$a = \left[1.175\left(\frac{Y_{50}}{T}\right)^{0.557}\right]^c$$ (C.21)

In solving Equation C.15 for use with this manual, it is assumed that the axes of a riprap element are approximately equal for standard riprap gradations. The mean diameter, D_{50}, is therefore substituted for Y_{50} and S_{50} parameters.

This page intentionally left blank.

APPENDIX D: RIPRAP STABILITY ON A STEEP SLOPE

The design of riprap for steep gradient channels presents special problems. On steep gradients, the riprap size required to stabilize the channel is often of the same order of magnitude as the depth of flow. The riprap elements often protrude from the flow, creating a very complex flow condition.

Laboratory studies and field measurements (Bathurst, 1985) of steep gradient channels have shown that additional factors need to be considered when computing hydraulic conditions and riprap stability. The development of design procedures for this manual has, therefore, been based on equations that are more general in nature and account directly for several additional forces affecting riprap stability. The design equation used for steep slopes in Chapter 6 of this manual is as follows. This appendix provides additional information on the development of the equation.

$$D_{50} \geq \frac{SF\, d\, S\, \Delta}{F_*(SG-1)} \tag{D.1}$$

where,

D_{50}	=	mean riprap size, m (ft)
SF	=	safety factor
d	=	maximum channel depth, m (ft)
S	=	channel slope, m/m (ft/ft)
Δ	=	function of channel geometry and riprap size
F_*	=	Shield's parameter, dimensionless
SG	=	specific gravity of rock (γ_s/γ), dimensionless

The stability of riprap is determined by analyzing the forces acting on individual riprap element and calculating the factor of safety against its movements. The forces acting on a riprap element are its weight (W_s), the drag force acting in the direction of flow (F_d), and the lift force acting to lift the particle off the bed (F_L). Figure D.1 illustrates an individual element and the forces acting on it.

The geometric terms required to completely describe the stability of a riprap element include:

α	=	angle of the channel bed slope
β	=	angle between the weight vector and the weight/drag resultant vector in the plane of the side slope
δ	=	angle between the drag vector and the weight/drag resultant vector in the plane of the side slope
θ	=	angle of the channel side slope
ϕ	=	angle of repose for the riprap

As the element will tend to roll rather than slide, its stability is analyzed by calculating the moments causing the particle to roll about the contact point, c, with an adjacent riprap element as shown in Figure D.1. The equation describing the equilibrium of the particle is:

$$\ell_2\, W_s\, \cos\theta = \ell_1\, W_s\, \sin\theta\, \cos\beta\ \ \ell_3\, F_d\, \cos\delta + \ell_4\, F_L \tag{D.2}$$

The factor of safety against movement is the ratio of moments resisting motion over the moments causing motion. This yields:

$$SF = \frac{\ell_2 W_s \cos\theta}{\ell_1 W_s \sin\theta \cos\beta + \ell_3 F_d \cos\delta + \ell_4 F_L}$$ (D.3)

where,

SF   =   Safety Factor

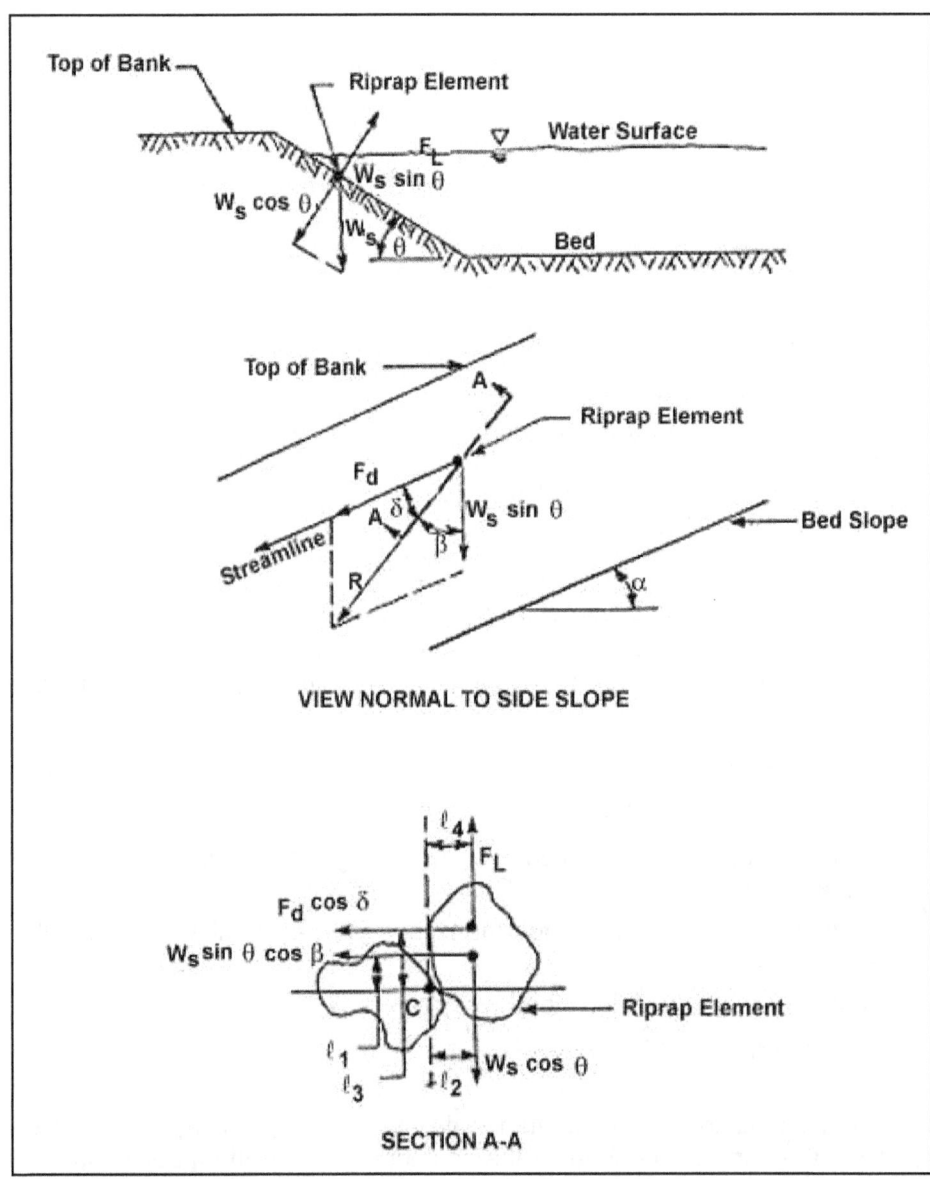

Figure D.1. Hydraulic Forces Acting on a Riprap Element

Evaluation of the forces and moment arms for Equation D.3 involves several assumptions and a complete derivation is given in Simons and Senturk (1977). The resulting set of equations are used to compute the factor of safety:

$$SF = \frac{\cos\theta \tan\phi}{\eta' \tan\phi + \sin\theta \cos\beta} \tag{D.4}$$

where,

η' = side slope stability number

The angles α and θ are determined directly from the channel slope and side slopes, respectively. Angle of repose, ϕ, may be obtained from Figure 6.1. Side slope stability number is defined as follows:

$$\eta' = \eta \frac{1 + \sin(\alpha + \beta)}{2} \tag{D.5}$$

where,

η = stability number

The stability number is a ratio of side slope shear stress to the riprap permissible shear stress as defined in the following equation:

$$\eta = \frac{\tau_s}{F_*(\gamma_s - \gamma)D_{50}} \tag{D.6}$$

where ,

τ_s = side slope shear stress = $K_1\tau_d$, N/m^2 (lb/ft^2)

F_* = dimensionless critical shear stress (Shields parameter)

γ_s = specific weight of the stone, N/m^3 (lb/ft^3)

γ = specific weight of water, N/m^3 (lb/ft^3)

D_{50} = median diameter of the riprap, m (ft)

Finally, β is defined by:

$$\beta = \tan^{-1}\left(\frac{\cos\alpha}{\dfrac{2\sin\theta}{\eta\tan\phi} + \sin\alpha}\right) \tag{D.7}$$

Returning to design Equation D.1, the parameter Δ can be defined by substituting equations D.5 and D.6 into Equation D.4 and solving for D_{50}. It follows that:

$$\Delta = \frac{K_1(1 + \sin(\alpha + \beta))\tan\phi}{2(\cos\theta \tan\phi - SF \sin\theta \cos\beta)} \tag{D.8}$$

Solving for D_{50} using Equations D.1 and D.8 is iterative because the D_{50} must be known to determine the flow depth and the angle β. These values are then used to solve for D_{50}. As discussed in Chapter 6, the appropriate values for Shields' parameter and Safety Factor are given in Table 6.1

APPENDIX E: FALL-BOARD TEST FOR GRASS DENSITY-STIFFNESS PARAMETER, C_S

The fall-board test is a simple way of estimating the combined effect of grass density and stiffness. The test should be conducted at random locations within an area with a homogeneous grass species mix and grass height. A standard board size and length is required. As shown in Figure E.1, the board is stood on end and allowed to freely rotate to the grass cover. When the board hits the grass, it slides length wise in the direction of rotation, imparting a friction force, which along with the weight of the board, deflects the grass in a manner that is similar to flowing water. The test should be repeated a sufficient number of times within an area to produce a reliable average value for the fall height, h_b.

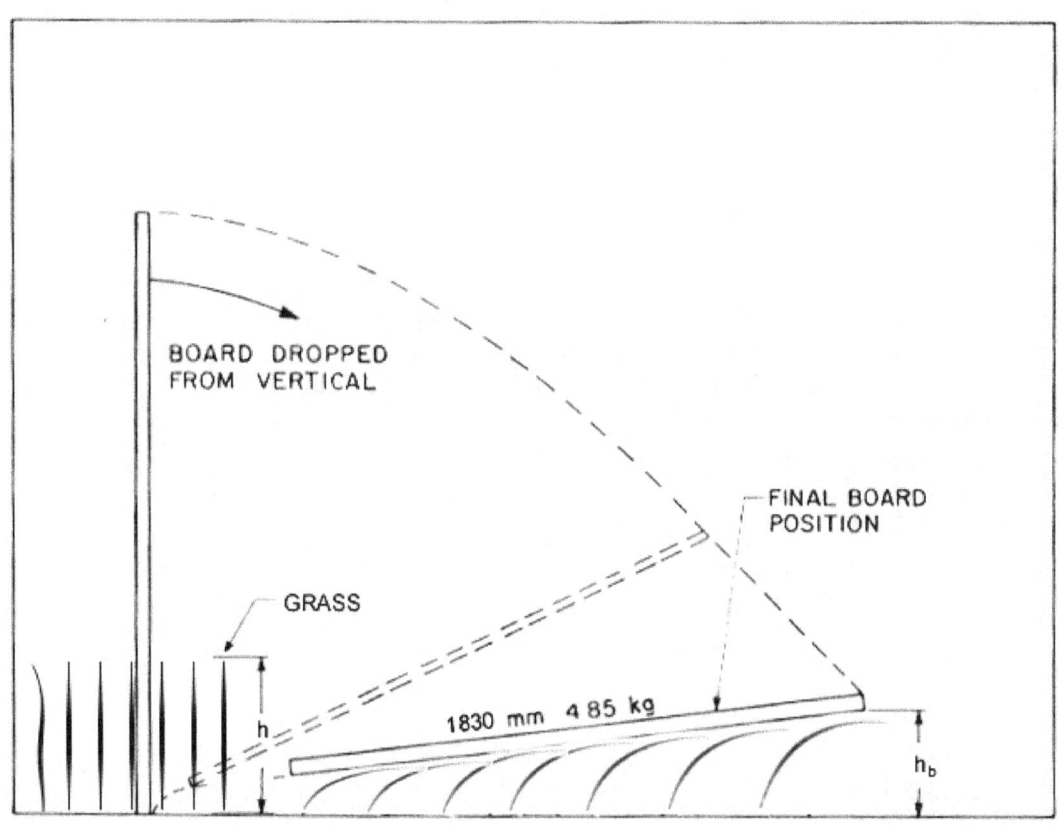

Figure E.1. Schematic of the Fall-board Test (after Kouwen, 1988)

Materials:

Wooden board	= 455 kg/m³ (28.5 lb/ft³) (i.e., ponderosa pine, redwood, spruce)
Board dimensions	= 1.829 m x 0.305 m (6 ft x 1 ft)
Board weight	= 4.85 kg (10.7 lb)

Test Procedure:

1. Select an area with homogeneous mix of grass species and grass height.

2. Randomly select locations within this area to conduct the fall-board test.

3. At each location record the height of grass stems, h.

E -1

4. Stand the board vertically on one end and freely allow it to fall over. As the board falls, it will rotate about the end in contact with the ground and slide forward.

5. Record fall height, h_b, as the distance between the ground and the bottom edge of the fallen end of the board.

6. Repeat the test at the other locations.

Calculations:

1. Compute the average of the grass height measurements and the fall height measurements.

2. Compute the grass density-stiffness coefficient using the following equation:

$$C_s = \alpha \left(\frac{h_b}{h} \right)^{2.82} \tag{E.1}$$

where,

C_s = Density-stiffness coefficient

α = unit conversion constant 3120 (SI), 265 (CU)

Design Example (SI)

A 100 m length of roadside ditch has a mature grass lining that was planted from a standard grass seed mix that is considered to be typical of that region of the state. Six (6) fall-board tests were conducted at random intervals along the ditch. Results of these tests are summarized in the following table.

Location (station)	0+018	0+025	0+037	0+053	0+075	0+093	**Mean**
Grass height, h (m)	0.159	0.158	0.153	0.154	0.143	0.147	**0.152**
Fall-board height, h_b (m)	0.043	0.034	0.049	0.037	0.030	0.036	**0.038**

$$C_s = \alpha \left(\frac{h_b}{h} \right)^{2.82} = 3120 \left(\frac{0.038}{0.147} \right)^{2.82} = 69$$

Design Example (CU)

A 300 ft length of roadside ditch has a mature grass lining that was planted from a standard grass seed mix that is considered to be typical of that region of the state. Six (6) fall-board tests were conducted at random intervals along the ditch. Results of these tests are summarized in the following table.

Location (station)	0+48	0+75	1+11	1+59	2+25	2+79	**Mean**
Grass height, h (ft)	0.52	0.52	0.50	0.51	0.47	0.48	**0.50**
Fall-board height, h_b (ft)	0.14	0.11	0.16	0.12	0.10	0.12	**0.13**

$$C_s = \alpha \left(\frac{h_b}{h} \right)^{2.82} = 265 \left(\frac{0.12}{0.48} \right)^{2.82} = 5.3$$

APPENDIX F: SHEAR STRESS RELATIONSHIP FOR RECPS

The general relationship for the transmission of shear to soil surface beneath a manufactured unvegetated lining is given by (Cotton, 1993; Gharabaghi, et al., 2002; Robeson, et al., 2003):

$$\tau_e = m(\tau_d - \tau_c) \tag{F.1}$$

where,

τ_e =   effective shear stress on the soil surface, N/m^2 (lb/ft^2)

m   =   rate of shear transmission

τ_d =   applied (design) shear at the surface of the lining, N/m^2 (lb/ft^2)

τ_c =   shear at which the soil surface is first mobile (i.e. critical shear), N/m^2 (lb/ft^2)

The critical shear value, τ_c, is for the soil type described in testing procedure ASTM D 6460.

Laboratory testing (McWhorter et al., 1968; Sanders, et al., 1990; Israelsen, et al., 1991; Northcutt, 1996; Robeson, et al. 2003) has shown that the rate of soil loss remains nearly constant and does not change with increasing shear stress over a wide range of applied shear stress. Data from Robeson, et al. (2003) is presented in Figure F.1 and shows this linear relationship for four lining types.

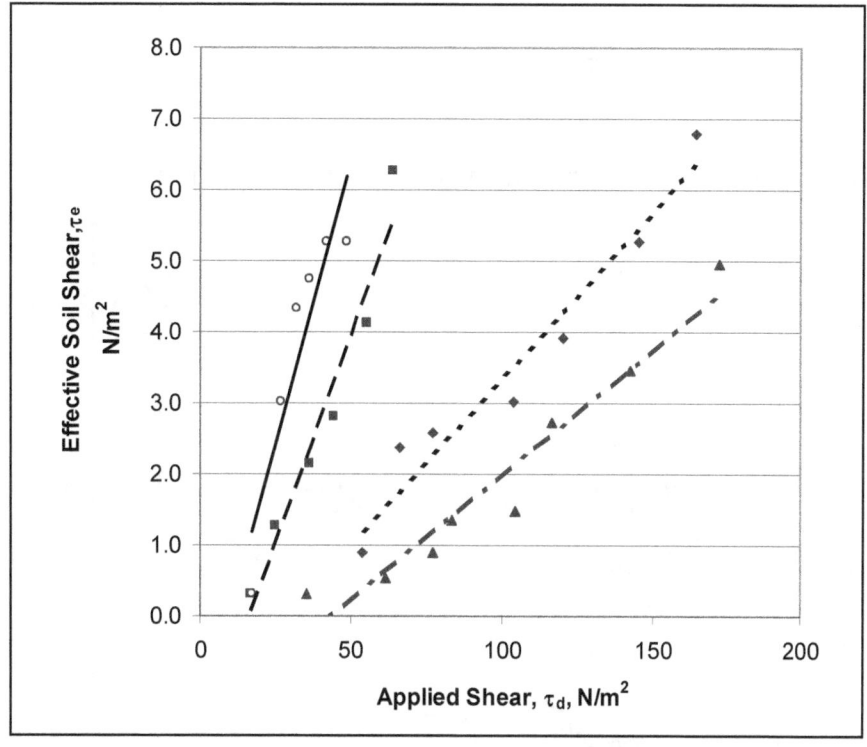

Figure F.1. Soil Shear versus Applied Shear to the Manufactured Lining

For a wide range of product types, the critical shear and the rate of shear transmission is directly related to the applied shear on the lining at cumulative soil erosion of 12.7 mm (0.5 in) over 30 minutes for a specified soil type in accordance with the ASTM D 6460. This shear value

F-1

is referred to as the lining shear, τ_l. (Note, the lining shear should be determined by testing under the same conditions as recommended by the manufacturer, i.e. stapling pattern, check slots, etc.)

As would be expected the critical shear on the lining is just a linear extrapolation from τ_l.

$$\tau_c = \frac{\tau_l}{4.3} \tag{F.2}$$

where,

τ_l = applied shear (lining shear) at a cumulative erosion of 12.7 mm (0.5 in), N/m^2 (lb/ft^2)

Likewise, the rate of shear transmission correlates to:

$$m = \frac{\alpha}{\tau_l} \tag{F.3}$$

where,

α = unit conversion constant, 6.5 (SI), 0.14 (CU)

Combining Equation F.2 and Equation F.3 with Equation F.1 gives:

$$\tau_e = \left(\tau_d - \frac{\tau_l}{4.3} \right) \frac{\alpha}{\tau_l} \tag{F.4}$$

Comparison to data presented in Robeson, et al. is shown in Figure F.2.

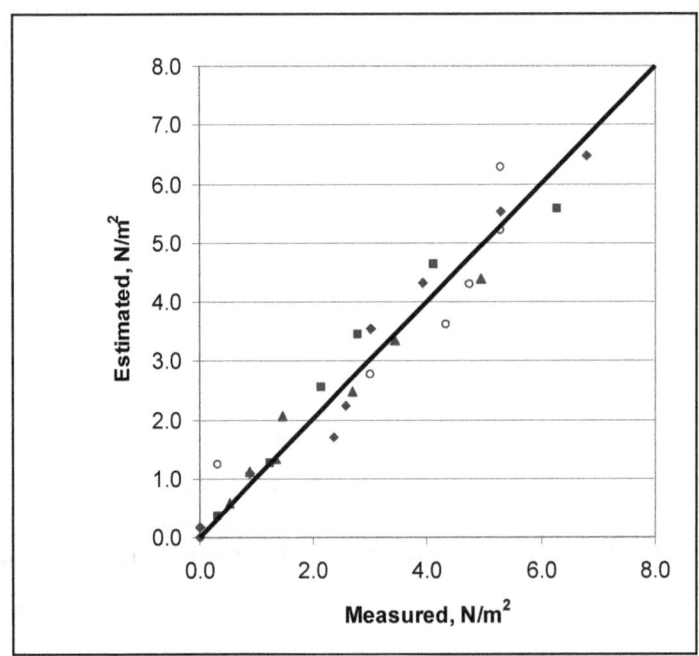

Figure F.2. Effective Shear on the Soil for Four RECP Linings

REFERENCES

AASHTO/NTPEP, 2002. "Rolled erosion control products (RECPs), Work Plan," National Transportation Product Evaluation Program (NTPEP), www.ntpep.org.

American Association of State Highway Officials, 1999. "Hydraulic Analysis and Design of Open Channels," Highway Drainage Guidelines, Volume VI.

Anderson, Alvin G., Amreek S. Paintal, and John T. Davenport, 1970. "Tentative Design Procedure for Riprap Lined Channels." NCHRP Report No. 108, Highway Research Board, National Academy of Sciences, Washington, D.C.

Bathurst, J.C., 1985. "Flow Resistance Estimation in Mountain Rivers," Journal of Hydraulic Engineering, ASCE, Vol. 111, No. 4.

Bathurst, J.C. R.M. Li, and D.B. Simons, 1981. "Resistance Equation for Large-Scale Roughness." Journal of the Hydraulics Division, ASCE, Vol. 107, No. HY12, Proc. Paper 14239, December, pp. 1593-1613.

Blodgett, J.C., 1986a. "Rock Riprap Design for Protection of Stream Channels Near Highway Structures, Volume 1 - Hydraulic Characteristics of Open Channels," USGS Water Resources Investigations Report 86-4127.

Blodgett, J.C., 1986b. "Rock Riprap Design for Protection of Stream Channels Near Highway Structures, Volume 2 - Evaluation of Riprap Design Procedures," USGS Water Resources Investigations Report 86-4128.

Blodgett, J.C. and C.E. McConaughy, 1985. "Evaluation of Design Practices for Rock Riprap Protection of Channels near Highway Structures," U.S. Geological Survey, Prepared in Cooperation with the Federal Highway Administration Preliminary Draft, Sacramento, California.

Chow, V.T. 1959. "Open Channel Hydraulics," New York, Company, McGraw-Hill Book

Clopper, Paul E., Dwight A Cabalka and Anthony G. Johnson, 1998. "Research, Development, and Implementation of Performance Testing Protocols for Channel Erosion Research Facilities (CERFS), International Erosion Control Association Proceedings of Conference XXIX.

Clopper, Paul E. and Yung-Hai Chen, 1988. "Minimizing Embankment Damage During Overtopping Flow," FHWA-RD-88-181, November.

Cotton, G.K., 1993. "Flow resistance properties of flexible linings," Hydraulic Engineering '93, Proceedings of American Society of Civil Engineering Conference, San Francisco, CA, pp 1344-1349.

Cox, R.L., R.C. Adams, and T.B. Lawson, 1971. "Erosion Control Study, Part II, Roadside Channels." Louisiana Department of Highways, in Cooperation with U.S. Department of Transportation, Federal Highway Administration.

Eastgate, W.I., 1966. "Vegetated Stabilization of Grasses, Waterways and Dam Bywashes," Masters Engineer Science Thesis, Department of Civil Engineering, University of Queensland, Australia.

Federal Highway Administration, 1983. "Hydraulic Design of Energy Dissipators for Culverts and Channels." Hydraulic Engineering Circular No. 14, FHWA-EPD-86-110.

Federal Highway Administration, 1989. "Design of Riprap Revetment," Hydraulic Engineering Circular No. 11, FHWA-IP-89-016.

Federal Highway Administration, 1998. "Geosynthetic Design and Construction Guidelines," FHWA-HI-95-038.

Federal Highway Administration, 2001. "Introduction to Highway Hydraulics," Hydraulic Design Series No. 4, FHWA-NHI-01-019.

Gessler, Johannes (1964). "The beginning of bed load movement of mixtures investigated as natural armoring in channels," Report Number 69 of the Laboratory of Hydraulic Research and Soil Mechanics of the Swiss Federal Institute of Technology.

Gharabaghi, B., W.T. Dickinson, and R.P. Ruda, 2002. "Evaluating Rolled Erosion Control Product Performance in Channel Applications", in Erosion Control January/February issue, Foster Publications.

Israelsen, C.E. and G. Urroz, 1991. "High velocity flow testing of turf reinforcement mats and other erosion control materials," Utah Water Research Laboratory, Utah State University, Logan, Utah.

Jarrett, R.D., 1984. "Hydraulics of High Gradient Streams," Journal of the Hydraulics Division, ASCE, Volume 110(11).

Kilgore, Roger T., and G. K. Young, 1993. "Riprap Incipient Motion and Shields' Parameter." Proceedings of the American Society of Civil Engineers' Hydraulics Division Conference.

Kouwen, N., 1988. "Field Estimation of the Biomechanical Properties of Grass," Journal of Hydraulic Research, 26(5), 559-568.

Kouwen, N. and R.M. Li, 1981. "Flow Resistance in Vegetated Waterways," Transactions of the American Society of Agricultural Engineering, 24(3), 684-690.

Kouwen, N., R.M. Li, and D.B. Simons, 1980. "Velocity Measurements in a Channel Lined with Flexible Plastic Roughness Elements." Technical Report No. CER79-80-RML-DBS-11, Department of Civil Engineering, Colorado State University, Fort Collins, Colorado.

Kouwen, N. and T.E. Unny, 1969. "Flexible Roughness in Open Channels," Journal of the Hydraulics Division, ASCE, 99(5), 713-728.

Kouwen, N., T.E. Unny, and H.M. Hill, 1969 "Flow Retardance in Vegetated Channel." Journal of the Irrigation and Drainage Division, IRE, pp. 329-342.

Lancaster, T., 1996. "A three-phase reinforced turf system, a new method for developing geosynthetically reinforced vegetated linings for permanent channel protection," International Erosion Control Association Proceedings of Conference XXVII, pp. 346-354.

Lane, E., 1955. "Design of Stable Channels," Transactions ASCE, Vol. 120.

Limerinos, J. T., 1970. "Determination of the Manning coefficient form measured bed roughness in natural channels," U.S. Geological Survey Water Supply Paper 1898-B, p. B1-B47.

Lipscomb, C.M., M.S. Theisen, C.I. Thornton, and S.R. Abt, 2003. "Performance Testing of Vegetated Systems and Engineered Vegetated Systems," International Erosion Control Association Proceedings of Conference XXXIV.

McWhorter, J. C., T.G. Carpenter, and R.N. Clark, 1968. "Erosion Control Criteria for Drainage Channels." Conducted for Mississippi State Highway Department in Cooperation with U.S. Department of Transportation, Federal Highway Administration, by the Agricultural and Biological Engineering Department, Agricultural Experiment Station, Mississippi State University, State College, Mississippi.

Northcutt, Paul E., 1996. "Final Performance Analysis – 1995 Evaluation Cycle 1 – Slope Protection Class 2 – Flexible Channel Liners," Texas Department of Transportation in cooperation with the Texas Transportation Institute.

Nouh, M.A. and R.D. Townsend, 1979. "Shear Stress Distribution in Stable Channel Bends." Journal of the Hydraulics Division, ASCE, Vol. 105, No. HY10, Proc. Paper 14598, October, pp. 1233-1245.

Olsen, O. J. and Q.L. Florey, 1952. "Sedimentation Studies in Open Channels Boundary Shear and Velocity Distribution by Membrane Analogy, Analytical and Finite-Difference Methods," reviewed by D. McHenry and R.E. Clover, U.S. Bureau of Reclamation, Laboratory Report N. SP-34, August 5.

Richardson, E.V., D.B. Simons and P.Y. Julien, 1990. "Highways in the River Environment." FHWA-HI-90-016.

Robeson, M.D., C.I. Thornton and R.J. Nelsen, 2003. "Comparison of results from bench and full scale hydraulic performance testing of unvegetated RECPs," International Erosion Control Association Proceedings of Conference XXXIV.

Sanders, T.G., S.R. Abt and P.E. Clopper, 1990. " A quantitative test of erosion control materials," International Erosion Control Association, XXI Annual Conference, Washington, D.C., p 209-212.

Santha, L. and C.R. Santha, 1995. "Effective utilization of fully matured grass in erosion control," International Erosion Control Association Proceedings of Conference XXVI, pp. 161-171.

Simons, D. B., Y. H. Chen, and L. J. Swenson, 1984. "Hydraulic Test to Develop Design Criteria for the Use of Reno Mattresses," Colorado Statue University, prepared for Maccaferri Steel Wire Products, Ltd., March.

Simons, D.B. and F. Senturk, 1977. "Sediment Transport Technology" Fort Collins, Colorado: Water Resources Publications.

Smerdon, E.T. and R.P. Beaseley, 1959. "The Tractive Force Theory Applied to Stability of Open Channels in Cohesive Soils." Agricultural Experiment Station, Research Bulletin No. 715, University of Missouri, Columbia, Missouri, October.

Soil Conservation Service, 1954. "Handbook of Channel Design for Soil and Water Conservation." SCS-TP-61, Stillwater, Oklahoma, revised.

Thibodeaux, K.G., "Performance of Temporary Ditch Linings." Interim Reports 1 to 17, Prepared for Federal Highway Administration by U.S. Geological Survey, Gulf Coast Hydroscience Center and Computer Science Corporation, 1982 to 1985.

U.S. Army Corps of Engineers, 1980. "Wire Mesh Gabions (Slope and Channel Protection)," CW-02541.

U.S. Army Corps of Engineers, 1994. "Hydraulic Design of Flood Control Channels," Engineer Manual 1110-2-1601, June.

U.S. Bureau of Reclamation, 1951. "Stable Channel Profiles," Lab. Report No. Hyd. 325, September 27.

U.S. Department of Agriculture, 1987. "Stability of grassed-lined open channels," Agricultural Research Service, Agricultural Handbook Number 667.

Wang, S.Y. and H.W. Shen, 1985. "Incipient Sediment Motion and Riprap Design." Journal of Hydraulics Division, ASCE, Vol. 111, No. 3, March, pp. 52-538.

Young, G. K., et al., 1996. "HYDRAIN - Integrated Drainage Design Computer System: Version 6.0 - Volume VI: HYCHL, FHWA-SA-96-064, June.